Lecture Notes in Mathematics

2100

For further volumes:
http://www.springer.com/series/304

Saint-Flour Probability Summer School

The Saint-Flour volumes are reflections of the courses given at the Saint-Flour Probability Summer School. Founded in 1971, this school is organised every year by the Laboratoire de Mathématiques (CNRS and Université Blaise Pascal, Clermont-Ferrand, France). It is intended for PhD students, teachers and researchers who are interested in probability theory, statistics, and in their applications.

The duration of each school is 13 days (it was 17 days up to 2005), and up to 70 participants can attend it. The aim is to provide, in three high-level courses, a comprehensive study of some fields in probability theory or Statistics. The lecturers are chosen by an international scientific board. The participants themselves also have the opportunity to give short lectures about their research work.

Participants are lodged and work in the same building, a former seminary built in the 18th century in the city of Saint-Flour, at an altitude of 900 m. The pleasant surroundings facilitate scientific discussion and exchange.

The Saint-Flour Probability Summer School is supported by:

– Université Blaise Pascal
– Centre National de la Recherche Scientifique (C.N.R.S.)
– Ministère délégué à l'Enseignement supérieur et à la Recherche

For more information, see back pages of the book and
http://math.univ-bpclermont.fr/stflour/

Jean Picard
Summer School Chairman
Laboratoire de Mathématiques
Université Blaise Pascal
63177 Aubière Cedex
France

Itai Benjamini

Coarse Geometry and Randomness

Ecole d'Été de Probabilités
de Saint-Flour XLI – 2011

 Springer

Itai Benjamini
Department of Mathematics
The Weizmann Institute of Science
Rehovot, Israel

ISBN 978-3-319-02575-9   ISBN 978-3-319-02576-6 (eBook)
DOI 10.1007/978-3-319-02576-6
Springer Cham Heidelberg New York Dordrecht London

Lecture Notes in Mathematics ISSN print edition: 0075-8434
                    ISSN electronic edition: 1617-9692

Library of Congress Control Number: 2013953863

Mathematics Subject Classification (2010): 82B43, 82B41, 05C81, 05C10, 05C80

Printed on acid-free paper

Springer is part of Springer Science+Business Media (www.springer.com)

Preface

The first part of the notes reviews several coarse geometric concepts. We will then move on and look at the manifestation of the underling geometry in the behavior of random processes, mostly percolation and random walk.

The study of the geometry of infinite vertex transitive graphs and Cayley graphs in particular is rather well developed. One goal of these notes is to point to some random metric spaces modeled by graphs that turn out to be somewhat exotic. That is, admitting a combination of properties not encountered in the vertex transitive world. These include percolation cluster on vertex transitive graphs, critical clusters, local and scaling limits of graphs, long range percolation, CCCP graphs obtained by contracting percolation clusters on graphs, and stationary random graphs including the uniform infinite planar triangulation (UIPT) and the stochastic hyperbolic planar quadrangulation.

Chapter 5 is due to Nicolas Curien, Chap. 12 was written by Ariel Yadin, and Chap. 13 is joint work with Gady Kozma.

I would like to deeply thank Omer Angel, Louigi Addario-Berry, Agelos Georgakopoulos, and Vladimir Shchur for comments, remarks, and corrections, and Nicolas Curien, Ron Rosenthal, and Johan Tykesson for *great help* with editing, collecting, and joining together the material presented.

Some of the proofs will only be sketched, or left as exercises to the reader. References to where proofs can be found in full detail are given throughout the text. Exercises and open problems can be found in most sections.

Excellent sources covering related material are Lyons with Peres [Lyo09], Pete [Pet09], Peres [Per99], and Woess [Woe05].

Thanks to N., Jean Picard and the St. Flour school organizers.

Contents

1 Introductory Graph and Metric Notions 1

2 On the Structure of Vertex Transitive Graphs 19

3 The Hyperbolic Plane and Hyperbolic Graphs 23

4 Percolation on Graphs 33

5 Local Limits of Graphs 41

6 Random Planar Geometry.. 53

7 Growth and Isoperimetric Profile of Planar Graphs 59

8 Critical Percolation on Non-Amenable Groups 63

9 Uniqueness of the Infinite Percolation Cluster 69

10 Percolation Perturbations 85

11 Percolation on Expanders 97

12 Harmonic Functions on Graphs 107

13 Nonamenable Liouville Graphs 121

References 125

Chapter 1
Introductory Graph and Metric Notions

In this section we start by reviewing some geometric properties of graphs. Those will be related to the behavior of random processes on the graphs in later sections.

A *graph* G is a couple $G = (V, E)$, where V denotes the set of vertices of G and E is its set of *undirected* edges. Throughout this lectures we assume (unless stated otherwise) that graphs are *simple*, that is, they do not contain multiple edges or loops. Thus elements of E will be written in the form $\{u, v\}$ where $u, v \in V$ are two different vertices. The degree of a vertex v in G, denoted $\deg(v)$, is the number of edges attached to (i.e. containing) v. We say that a graph G has a bounded degree if there exists a constant $M > 0$ such that $\deg(v) < M$ for all vertices $v \in V$. Our graphs may be finite or infinite, however, in what follows (except when explicitly mentioned) all the graphs considered are assumed to be countable and locally finite, i.e. all of their vertices are of finite degree. The graph distance in G is defined by

$$d_G(v, w) = \text{the length of a minimal path from } v \text{ to } w.$$

A graph is *vertex transitive* if for any pair of vertices in the graph, there is a graph automorphism mapping one to the other, formally,

Definition 1.1.

1. Let $G = (V, E)$ be a graph. A bijection $g : V \to V$ such that $\{g(u), g(v)\} \in E$ if and only if $\{u, v\} \in E$ is called a *graph automorphism*. The set of all automorphisms of G is denoted by $\mathrm{Aut}(G)$.
2. A graph G is called vertex transitive if for every $u, v \in V$ there exists a graph automorphism mapping u to v.

Recall, *simple random walk* (SRW) on a graph, is a (discrete time) random walk that picks the next state uniformly over the neighbors of the current state.

I. Benjamini, *Coarse 3Geometry and Randomness*, Lecture Notes in Mathematics 2100, DOI 10.1007/978-3-319-02576-6__1,
© Springer International Publishing Switzerland 2013

1.1 The Cheeger Constant

Isoperimetric problems were known already to the ancient Greeks. The first problem on record was to design a port, which was reduced to the problem of finding a region of maximal possible area bounded by a given straight line and a curve of a prescribed length whose endpoints belong to the line. The solution is of course the semi-circle.

Given $S \subset V$, we define the *outer boundary* of S to be

$$\partial S := \{u \notin S \ : \ \exists v \in S \text{ such that } \{u, v\} \in E\}.$$

Thus ∂S contains all neighbors of elements in S that are not themselves in S. Note that there are other similar notions for the boundary of a set in a graph (the inner boundary, and the edge boundary). In many situations these may be used as well. We are interested in the ratio between the size of a set and the size of its boundary. This leads to the following definition:

Definition 1.2. Let $G = (V, E)$ be a finite graph. The Cheeger constant of G is defined to be

$$h(G) = \inf \left\{ \frac{|\partial S|}{|S|} \ : \ S \subset V \, , \, 0 < |S| \le \frac{|V|}{2} \right\}.$$

If G is an infinite graph the Cheeger constant is defined in the same manner by

$$h(G) = \inf \left\{ \frac{|\partial S|}{|S|} \ : \ S \subset V \, , \, 0 < |S| < \infty \right\}.$$

An infinite graph G with $h(G) > 0$ is called non-amenable and amenable otherwise.

Exercise 1.3. (Level 1) Here are some examples, given as exercises, for the value of the Cheeger constant:

1. Show that $h(\mathbb{Z}^d) = 0$ for all $d \in \mathbb{N}$.
2. Show that $h(\text{Binary Tree}) = 1$.

One reason for which the Cheeger constant is interesting is because it is an indicator for bottle necks in the graph. See Exercise 4.12 and Fig. 1.1

There are many versions for the Cheeger constant, one such version is the following

Definition 1.4. The *Anchored Expansion* or *Rooted Isoperimetric Constant* of an infinite graph G in a vertex $v \in V$ is

$$\text{Ah}_v(G) = \inf_{v \in S} \frac{|\partial S|}{|S|},$$

where the infimum is taken over all finite connected subsets of vertices containing the vertex v.

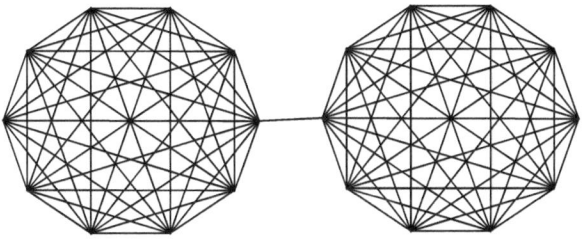

Fig. 1.1 A connected graph with a bottle neck

Exercise 1.5. (Level 1) Prove that for an infinite connected graph G, the statement $Ah_v(G) > 0$ is independent of the choice of v.

Exercise 1.6. (Level 3) Show that there is a connected graph G with $h(G) = 0$ but $Ah(G) > 0$.

Here is a first example for the connection between isoperimetric properties of a graph and the behavior of a random walk on it.

Theorem 1.7 ([Vir00]). *If G is an infinite connected graph with bounded degree such that $Ah_v(G) > 0$ then:*

1. *The Simple Random Walk (SRW) on G has a positive speed.*
2. *The probability of return to the origin of the simple random after n steps decays faster than $\exp(-cn^{1/3})$ for some constant $c > 0$.*

1.2 Expander Graphs

Definition 1.8. A graph G is said to be d-regular if each vertex $v \in V$ has degree d. A family of d-regular finite connected graphs $\{G_n\}_{n=1}^{\infty}$ is said to be an expander family if the following two conditions hold:

1. $|V(G_n)|$ is strictly increasing.
2. There exists a constant $C > 0$ such that $h(G_n) > C$ for every $n \in \mathbb{N}$.

Note that we require d to be fixed, i.e., we don't allow it to depend on n. For example, we don't think of $\{K_n\}_{n=1}^{\infty}$ (complete graphs on n vertices) as an expander family.

Example 1.9. For every $n \in \mathbb{N}$ choose a random d-regular graph uniformly from the set of all d-regular graphs with n vertices. It is a well known fact that the resulting graph sequence has a large Cheeger constant with high probability (see for example [HLW06]). Even though this gives a probabilistic method for constructing an expander, constructing an explicit family of expanders is not a simple task and has been the focus of extensive research in the past three decades.

The inequality in the expander definition is an isoperimetric inequality (when we think of the set size as its area, and the boundary size as its perimeter). This inequality implies that the graph has very strong connectivity properties and has many applications [HLW06].

Exercise 1.10 ("The Music of Chance"). (Level 3) Consider the n-cycle C_n with $n \in \mathbb{N}$ even. We create a new graph via the following procedure: Choose two different vertices of C_n uniformly and add an edge between them. Next choose two additional different vertices (from the set of vertices not to be chosen yet) and add an edge between them. Continue in this fashion until all vertices are chosen. Denote the graph obtained in the last process by G_n. Prove that there exist constants $c, p > 0$ such that for every n, $\mathbb{P}(h(G_n) > c) > p$.

Open problem 1.11. *Is there an infinite bounded degree graph in which all balls are (uniform) expanders? That is, there is some $c > 0$, so that for any ball[1] $B = B(v, r)$ in G we have $h(B) > c$?*

We conjecture that the answer is no.

Exercise 1.12. (Level 2) Construct an infinite graph with the property that all balls with some fixed center are expanders.

1.3   Isoperimetric Dimension

The Cheeger constant of $\mathbb{Z}^d$ is zero for every d and so it doesn't help us to distinguish between the Euclidean lattices. In order to study the difference between them we define a new parameter for graphs: *the isoperimetric dimension*:

Definition 1.13. The isoperimetric dimension of an infinite connected graph G is

$$\mathrm{I} - \dim(G) := \sup \left\{ d \geq 1 : \exists c > 0, \forall S \subset V \text{ with } 0 < |S| < \infty, |\partial S| \geq c|S|^{\frac{d-1}{d}} \right\}.$$

In words, if $\mathrm{I} - \dim(G) = d$, then for every set S of size n, the boundary of S is at least of order $n^{\frac{d-1}{d}}$ up to some constant factor independent of n, and d is the largest number for which this property holds.

Exercise 1.14. (Level 2) Prove that $\mathrm{I} - \dim(\mathbb{Z}^d) = d$. See [BL90] for a stronger result on the torus.

Exercise 1.15. (Level 3) Construct a graph with $\mathrm{I} - \dim(G) = \infty$ and $h(G) = 0$.

Exercise 1.16. (Level 2) Show that any $r \geq 1$ can appear as the isoperimetric dimension.

---

[1] For every choice of $v \in V$ and $r > 0$.

Exercise 1.17. (Level 4) Construct a graph $G = (V, E)$ for which there exists a constant $c > 0$ such that $|B(v, r)| < c r^\beta$ for every $v \in V$ and every $r > 0$ and such that for any set of size n the boundary of the set is of size bigger than $c' n^{\beta-1}$. See [Laa00] for more details.

Here is a generalization of the isoperimetric dimension:

Definition 1.18. The *isometric profile* of a graph $G = (V, E)$ is a function $f : \mathbb{N} \to [0, \infty)$ defined by

$$f(n) = \inf\{|\partial S| : |S| \le n, \ S \subseteq G\}.$$

In [Koz07] Gady Kozma proved that for a planar graph[2] of polynomial growth[3] isoperimetric dimension strictly bigger than 1 implies that $p_c < 1$. Here p_c is the critical value of bond percolation process on the graph (for a precise definition see Chap. 2). For general graphs this question is still open. It is a classic result that knowledge on the Isoperimetric dimension implies upper bounds on the return probability of random walks on graphs, see [VSCC92].

Theorem 1.19 ([Bow95]). *Let G be an infinite connected planar graph then exactly one of the following holds:*

- $h(G) > 0$.
- $\mathrm{I} - \dim(G) \le 2$.

Exercise 1.20. (Level 3) Prove Theorem 1.19 for planar triangulations of degree bigger than or equal to 6.

1.4 Separation

Here we define one last geometric parameter for graphs known as the Separation Profile. We start with the following definition:

Definition 1.21. Let G be a finite graph of size n. We denote by $\mathrm{Cut}(G)$ the minimal number of vertices needed to be removed from G in order to create a new graph whose largest connected component is smaller than $\frac{n}{2}$. (Removing a vertex also removes any edge containing it.)

Theorem 1.22 ([LT79]). *For every planar graph G of size n, $\mathrm{Cut}(G) \le C \sqrt{n}$, where C is some universal constant. In fact, this also holds for graphs which can be embedded in any fixed genus surface.*

---

[2] A graph that can be embedded in the plane.

[3] The number of vertices at distance $\le n$ from some fixed vertex growth polynomially.

Definition 1.23. Let G be a graph. The *separation profile* of G is a function $g :$ $\mathbb{N} \to [0, \infty)$ given by

$$g(n) = \sup_{|H|=n} \mathrm{Cut}(H),$$

where the supremum is taken over all subgraphs H of G of appropriate size.

Exercise 1.24. (Level 3) What is the separation profile of $T \times \mathbb{Z}$, where T is the 3 regular tree?

For the separation profile $T \times T$, and more on separation see [BST12], which includes many open questions, and [MTTV98].

1.5   Rough Isometry

If two metric spaces are isometric (i.e. there is a metric preserving bijection between them) then in a sense they are the same space. It is very useful to consider a weaker sense of isometry in which two metric spaces may be similar. The notion of *rough isometry* captures several important properties of metric spaces. Sometimes it is called *quasi isometry*.

Definition 1.25. Let H and G be two metric spaces with corresponding metrics d_H and d_G. We say that H is *rough isometric* to G if there is a function $f : G \to H$ and some constant $0 < C < \infty$ such that

(i) For all $x, y \in G$

$$C^{-1} \cdot d_H(f(x), f(y)) - C \leq d_G(x, y) \leq C \cdot d_H(f(x), f(y)) + C.$$

(ii) The range of f is a C-net in H, that is, for every $y \in H$ there is $x \in G$ so that

$$d_H(f(x), y) < C.$$

Exercise 1.26. (Level 1) Prove that rough isometry is an equivalence relation.

Exercise 1.27. (Level 1) Show that the spaces $\mathbb{R}^2$, $\mathbb{Z}^2$ and the two dimensional triangular lattice are roughly isometric to each other.

Another useful coarse geometric notion is that of *almost planarity*. A graph is k-almost planar if it can be drown in the plan so that each edge is crossed at most k times.

Exercise 1.28. (Level 2) Give an example of an almost planar graph which is not rough isometric to a planar graph.

Exercise 1.29. (Level 2) Prove that $\mathbb{Z}^d$ is not roughly isometric to $\mathbb{Z}^k$ for any $k \neq d$. Similarly, prove that $\mathbb{R}^d$ is not roughly isometric to $\mathbb{R}^k$ for any $k \neq d$. *Hint:* use volume considerations.

Exercise 1.30. (Level 1) Prove that for any $d \in \mathbb{N}$ any two d-dimensional Banach spaces are roughly isometric.

Exercise 1.31. (Level 2) Is it true that all finite graphs are roughly isometric.

Open problem 1.32 (Gady Kozma's Question). *Is there a bounded degree (or even unbounded degree) graph G, which is roughly isometric to $\mathbb{R}^2$ (with the Euclidean metric) with multiplicative constant 1? (That is, $|\|f(x) - f(y)\|_2 - d_G(x, y)| < C$ for some $f : G \to \mathbb{R}^2$.) Note that $\mathbb{Z}^2$ is roughly isometric to $\mathbb{R}^2$ with multiplicative constant $\sqrt{2}$.*

The following exercise assumes basic knowledge of notions considered in the coming probabilistic sections.

Exercise 1.33. (Level 4) Which of the following properties are rough-isometry invariant? (The last two are still open.)

1. $p_c(G) < 1$.
2. Recurrence of SRW on the graph.
3. Liouville property, i.e. non existence of non constant bounded harmonic functions.
4. $p_c < p_u$.

Exercise 1.34. (Level 3) Show that a tree in which all degrees are 3 or 4 is roughly isometric to the 3 regular tree.

Exercise 1.35. (Level 4) Is the $\mathbb{Z}^2$ grid roughly isometric to a half plane?

Exercise 1.36. (Level 4) Is the binary tree roughly isometric to a vertex transitive graph?

Exercise 1.37. (Level 3) Give an example of a graph which is not roughly isometric to a vertex transitive graph.

Exercise 1.38. (Level 4) Show that a vertex transitive graphs with linear growth are roughly isometric to the line.

Exercise 1.39. (Level 4) Show that the 3-regular tree and the 4-regular trees are rough isometric.

Exercise 1.40. (Level 4) Show that T, $T \times \mathbb{Z}$, $T \times \mathbb{Z}^2$, $T \times T$ and $T \times T \times T$ are all not rough isometric.

1.6 Ends

In this subsection we consider the notion of graph ends.

Definition 1.41. Let $G = (V, E)$ be a connected graph and fix some $v \in V$. Denote by $k(r)$ the number of infinite connected components of the subgraph of

G induced by removing from it the ball $B(v, r)$. The number of *ends* of G is defined as $\lim_{r \to \infty} k(r)$.

Exercise 1.42. (Level 2) Show that for connected graphs, the number of ends is independent of the choice of the vertex v used in the definition and that it is well defined.

We recall our previous definition (Definition 9.29) for ends in a graph:

Definition 1.43. A ray in an infinite graph is an (semi) infinite simple path. Two rays are said to be equivalent if there is a third ray (which is not necessarily different from either of the first two rays) that contains infinitely many of the vertices from both rays. This is an equivalence relation. The equivalence classes are called ends of the graph.

Exercise 1.44. (Level 3) Show that the number of ends of a graph under the two definitions is the same.

When a graph has infinitely many ends, this also allow us to talk about the cardinality of the set of ends, and distinguish between graphs with countably or uncountably many ends.

Example 1.45.

- The graph $\mathbb{Z}$ has two ends.
- For $d > 1$, $\mathbb{Z}^d$ has one end.
- The d-regular tree has uncountably many ends.
- The comb graph consists of a copy of $\mathbb{Z}$ with an infinite path attached at each vertex. The comb has countably many ends.
- The $1 - 3$ *tree* has 2^ℓ vertices in the ℓth level. (when drawn in the plane) The left half of the vertices in level ℓ have one child in level $\ell + 1$, and the right half have three children in level $\ell + 1$. (Except at level 0 where the root has two children). The $1 - 3$ tree has countably many edges. Since every end (except one) is isomorphic to $\mathbb{Z}_+$ (when removing a large enough ball), the critical probability for percolation on the $1 - 3$ tree is $p_c = 1$.

Exercise 1.46. (Level 3) Show that the cardinality of the set of graph ends is invariant with respect to rough isometries for connected graphs.

Exercise 1.47. (Level 2) Let G be a vertex transitive graph with 2 ends. Prove that G is roughly isometric to $\mathbb{Z}$.

Exercise 1.48. (Level 3) Let G be a connected, vertex transitive graph with linear growth (that is, there exists $c < \infty$ such that $|B(v, t)| < ct$ for every $v \in V$ and $t > 0$). Prove that G is roughly isometric to $\mathbb{Z}$ or $\mathbb{N}$.

Theorem 1.49. *The number of ends of any infinite vertex transitive graph is either 1, 2, or ∞.*

See e.g. [Mei08] for a proof. See also Theorem 5.22 for a closely related result in the case of unimodular random graphs.

1.7   Graph Growth and the Cheeger Constant

In this subsection we consider the volume growth of graphs and the connection between uniform volume growth (to be defined) and the Cheeger constant. All the theorems in this section are stated without proofs. See the suggested references for additional information.

Definition 1.50. An infinite connected graph G is said to have the *uniform volume growth* if there exists a constant $C > 0$ so that for any choice of vertices $u, v \in G$ and any $r > 0$ we have $|B(v, r)| \le C|B(u, r)|$.

Example 1.51. A transitive graph has uniform volume growth with constant $C = 1$.

It is easy to see that if $h(G) > 0$ then G must have exponential growth, i.e. there are $C > 1$ and $K > 0$ so that $|B(v, r)| > K \cdot C^r$ for every $v \in G$ and $r > 0$. In fact, one can ensure that $C \ge 1 + h(G)$. This raises the interesting question of whether the converse is also true, namely, if G has exponential growth, does the Cheeger constant of G must be positive? If the graph is not assumed to be transitive, the answer is easily seen to be negative, for example one can take any graph with exponential volume growth, and attach an infinite path to some vertex. Even though the answer to the last question is negative the following theorem shows that the Cheeger constant cannot decrease to fast with the size of the sets:

Theorem 1.52 ([BS04]). *Let G be a graph with uniform exponential growth, then there exists some constant $c > 0$ such that for any set $S \subset G$ of size n the following isoperimetric inequality holds:*

$$\frac{|\partial S|}{|S|} \ge \frac{c}{\log n}.$$

In particular the graph G is transient

The following lemma shows that the answer to the question is negative even if one assumes that the graph is transitive or even a Cayley graph.

Lemma 1.53. *The lamplighter graph $LL(\mathbb{Z})$ has uniform exponential growth but $h(LL(\mathbb{Z})) = 0$.*

For definition of the lamplighter graph see Example 1.81 in Sect. 1.9.

Proof. First observe that the lamplighter graph has a uniform volume growth since it is vertex transitive. Consider all paths where the lamp-lighter moves $n/2$ steps to the right, and either flips or leaves unchanged the lamp at each intermediate site. There are $2^{\frac{n}{2}}$ such paths, each with length less than n, and hence $|B(v, n)| > 2^{\frac{n}{2}}$ (at least for even n), giving the exponential growth.

To see that $h(LL(\mathbb{Z})) = 0$, consider the set S_n consisting of all $(w, k) \in LL(\mathbb{Z})$ where $k \in [1, .., n]$ and the support of w is a subset of $[1, .., n]$. Then $|S_n| = n2^n$, but

$|\partial S_n| = 2 \cdot 2^n$, since the only vertices outside of S_n who are neighbors of vertices in S_n involve moving the walker to 0 or to $n + 1$. Thus

$$h(LL(\mathbb{Z})) \leq \frac{|\partial S_n|}{|S_n|} = \frac{1}{n} \xrightarrow[n\to\infty]{} 0.$$

□

Exercise 1.54. (Level 2) Show that in $LL(\mathbb{Z})$, the volume growth of a ball of radius R is α^R where $\alpha = \frac{1+\sqrt{5}}{2}$ (up to polynomial factors).

Exercise 1.55. (Level 2) Show that there exists a subgraph $H \subset LL(\mathbb{Z})$ such that $h(H) > 0$.

The following is about 30 years old open problem (as for 2011) which generalize the previous question:

Open problem 1.56. *Let G be vertex transitive graph of exponential growth. Is there always an infinite subgraph (a tree say) $H \subset G$ with $h(H) > 0$?*

If the answer to the last question is positive one can also ask the following:

Open problem 1.57. *Can one prove (or disprove) the above open problem when replacing the assumption of vertex transitivity by uniform exponential growth?*

Exponential growth on its own (without the assumption of transitivity or uniform growth) is not sufficient to imply existence of a subgraph with positive Cheeger constant. This is demonstrated by the following example.

Example 1.58. The canopy tree is defined as follows: Start with a copy of $\mathbb{N} = \{0, 1, \ldots\}$. For each $i > 0$, take a binary tree of depth $i - 1$, and attach its root to $i \in \mathbb{N}$ by an additional edge. It is easy to see that $|B(v, r)| > 2^{\frac{r}{2}}$ for any vertex v in the canopy graph and any $r > 0$. However, any finite set can be disconnected from infinity by removing a single vertex, and this property is shared by any subgraph H. Thus $h(H) = 0$ for any infinite subgraph.

Note that in the canopy tree some balls have volume of order $2^{\frac{r}{2}}$, while others have volume of order 2^r, and thus it does not have uniform volume growth. This graph is also recurrent.

We also have the following theorem regarding graphs with uniform exponential growth:

Theorem 1.59 ([BS04]). *Let G be a graph with a uniform exponential volume growth. Then G is transient with respect to the simple random walk.*

Arbitrary graphs can have any growth rate, but transitive graphs seem much more restricted. Polynomial and exponential growths are common, and so in 1965 John Milnor asked if there is a Cayley graph (or vertex transitive graph) of intermediate growth, i.e. with some vertex v so that $|B(v, r)|$ grows faster than any polynomial, but slower than any exponential. In 1985 Grigorchuk solved this problem in the affirmative and constructed a group with growth $|B(v,r)| \approx \exp(r^\alpha)$ where $\frac{1}{2} < \alpha < 1$. The question remains open for $0 < \alpha \leq \frac{1}{2}$. See [Nek05] for additional information.

The following example shows that without the constrain of uniform growth there are examples of such graphs G.

Example 1.60. Define a graph G by taking the graph $\mathbb{Z}$, and adding edges $\{n.n+2^k\}$ for every pair $k \in \mathbb{N}$ and $n \in \mathbb{Z}$ so that $2^{k-1}|n$ but $2^k \nmid n$.

With Oded we showed that if $h(G) > 0$ then G contains a subtree T with $h(T) > 0$.

Exercise 1.61. (Level 4) Assume G is transient. Does G contain a transient subtree?

Open problem 1.62. *Show that bounded degree transient hyperbolic graph contains a transient subtree. Bonk Schramm [BS00] might be useful.*

1.8 Scale Invariant Graphs

Recall that a δ-net in a metric space (X, d) is a set S such that for any $x \in X$, $d(x, S) \leq \delta$. Given a graph G, generate a k-net graph in G, by picking a maximal set of vertices, with respect to inclusion, such that the distance between any two vertices is at least k. Given a k-net in a graph G we can construct a new graph on the net by placing an edge between any two vertices in the net at distance at most $2k$. Let G_k denote any of the possible resulting k-net graphs of G. For any fixed k, it is easy to see that G_k is roughly isometric to G, with some constant C_k. We call G *scale invariant* if as k grows G_k stays roughly isometric to G with constants bounded away from 1 and ∞, i.e.

$$1 < \liminf C_k \leq \limsup C_k < \infty.$$

We call G *super scale invariant* if $C_k \to \infty$ as $k \to \infty$ and *sub scale invariant* if $C_k \to 1$ as $k \to \infty$.

See the paper by Nekrashevych and Pete [NP11] for additional information on the subject.

Exercise 1.63. (Level 2) Show that every graph falls into one of the three categories, and that this definition is independent of the choice of the k-nets G_k.

Exercise 1.64. (Level 2) Prove that the regular tree T_d is super scale invariant. More generally, any non-amenable graph is super scale invariant.

Example 1.65. Euclidean lattices, i.e. $\mathbb{Z}^d$, are scale invariant. More generally, any vertex transitive graph of polynomial growth is scale invariant.

Exercise 1.66. (Level 3) Let G be a vertex transitive graph. Show that G is scale invariant if and only if G satisfies the volume doubling property (i.e. there exists a constant $C > 0$ such that for all $v \in G$ and for all $r > 0$, $|B(x, 2r)| \leq C|B(x, r)|$)and therefore every vertex transitive which is scale invariant has polynomial growth.

Next we state three open problems regarding scale invariance of graphs.

Open problem 1.67. *Is there any vertex transitive sub scale invariant graph?*

Open problem 1.68. *If G is a Cayley graph of a finite group times $\mathbb{Z}$, then G_k for sufficiently large k is $\mathbb{Z}$ (note that there may be edges from n to $(n+2)$). Is there a graph that by keep passing to k-nets you get smaller and smaller graphs unboundedly many times?*

In general, rough isometry preserves the type of the growth rate, and also inequalities of the form $|\partial A| \leq C|A|^x$. A k-net graph is roughly isometric to the original graph, and therefore G_k has the same growth exponents and isoperimetric dimension as G. Similarly, rough isometry preserves recurrence of random walks.

Open problem 1.69. *Suppose G is vertex transitive. Can one show that $h(G_k) \geq h(G)$, as long as G_k is not one vertex?*

This question is non-trivial also for general (non transitive) graphs. In that case, the answer is negative, following from a counterexamples due to Oded Schramm (personal communication).

Example 1.70. Start with some graph having a Cheeger constant close to 1. We attach to it a finite structure which will not decrease the Cheeger constant, but will become a simple path in G_k for a large k. Thus G_k will have a small Cheeger constant. Pick some $n \gg k$. For $j \leq n$, Let H_j be the complete graph on 2^j vertices. Connect every vertex in H_j to every vertex in H_{j+1}. Finally, attach the vertices in H_n to some vertex in a graph G_0 with Cheeger constant 1. The resulting graph has bounded degrees (since we stopped at some finite n) and Cheeger constant close to 1. A k-net of the graph consists of a k-net of G_0, together with a path of length at least $\frac{n}{2k} - 1$. Thus it has Cheeger constant at most $\frac{2k}{n}$.

A weaker form of Question 1.69 is whether $h(G_k) > Ch(G)$ for some constant, possibly depending on the maximal degree of G. Without the assumption of transitivity, the answer is still negative, as showed by the following example:

Example 1.71. Take an infinite binary tree T with root o. Pick some large n. Let T' be an 8-ary tree of depth n. Let ϕ be the obvious bijection between the leaves of T' and the vertices at level $3n$ in T. Connect each leaf $v \in T'$ to $\phi(v)$ by an edge, and let G be the resulting graph. This graph has a k-net with Cheeger constant which is going to zero with n.

Here is another open problem:

Open problem 1.72. *From the rough-isometry invariance one can prove that there exists a positive function $f(h, d, k)$ such that if G has Cheeger constant*

$h(G) > h > 0$ *and maximal degree* $d_{\max} < d$, *then the* k-*net of* G *has Cheeger constant* $h(G_k) > f(h, d, k)$. *Can one show that* $\inf_k f(1, 10, k) > 0$. *Is it the case for* k-*nets in a regular tree? We believe that on a regular tree* $f(1, 10, k)$ *goes to infinity with* k *(and rather fast too).*

In [Pel10] Peled Studied rough isometries between random spaces. One can ask for a modification of the definition of scale invariance so that super critical percolation on $\mathbb{Z}^d$ or other random spaces can be considered scale invariant?

A natural way to weaken the notion of scale invariant is to consider a distribution on metric spaces or graphs, which is uniformly (roughly)-stationary under moving to a k-net. The family tree of critical Galton Watson process conditioned to survive is scale stationary. Are supercritical Galton Watson trees scale stationary? It follows from the work of Le Gall and Miermont, see [Gal11] and [Mie13], that the uniform infinite planar quadrangulation is a scale stationary distribution.

Given a distribution on graphs which is roughly scale stationary. Is there an exact scale stationary distribution (that is, not up to rough isometries) which is coupled to the original distribution, each coupled spaces are uniformly rough isometric?

A scale invariance problem, related to scaling limit of random planar metric, considered in Chap. 6. Start with a unit square divide it to four squares and now recursively at each stage pick a square uniformly at random from the current squares (ignoring their sizes) and divide it to four squares and so on.

Look at the minimal number of squares needed in order to connect the bottom left and top right corner with a connected set of squares.

Open problem 1.73. *Is there is a deterministic scaling function, such that after dividing the random minimal number of squares needed after n subdivisions by it, the result is a non degenerate random variable.*

We conjecture that the answer is yes. Does geodesic stabilize, as we further divide?

1.9 Examples of Graphs

This section introduces some useful examples of graphs, many more examples scattered along the notes. Mostly we will consider here graphs generated from finitely generated groups. Those graphs known as Cayley graphs give a geometric presentation of the group which can help in their study. There are many books on geometric group theory, see for example [Mei08] and [Pet09]. We start by recalling some definitions:

A countable group Γ is said to be finitely generated if there is a finite set of elements, $g_1, \ldots, g_k \in \Gamma$, so that any $\gamma \in \Gamma$ can be written as a finite product of those elements.

Example 1.74. The group $(\mathbb{Z}, +)$ is finitely generated by the set $\{-1, +1\}$.

Fig. 1.2 The grandfather graph. *Black edges* belong to the original tree and *red edges* are added according to the chosen direction

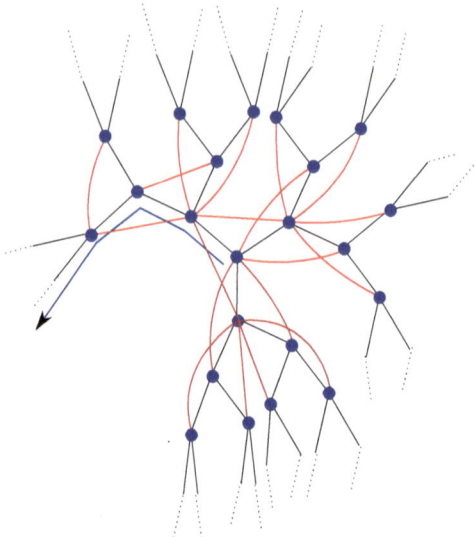

Example 1.75. F_2—the free group of two symbols. Take the symbols $\{a, b, a^{-1}, b^{-1}\}$. The identity of the group is e and all other elements are words generated by the 4 symbols.

Definition 1.76. Let Γ be a finitely generated group. The Cayley graph of Γ with respect to the generating set $S = \{g_i^{\pm 1}\}$, denoted $G_\Gamma = (V_\Gamma, E_\Gamma)$, is the graph with vertex set $V_G = \Gamma$, and edges

$$E_G = \{(\gamma_1, \gamma_2) \ : \ \gamma_1, \gamma_2 \in \Gamma, \exists g \in S \ s.t. \ \gamma_1 = g\gamma_2\}.$$

Exercise 1.77. (Level 1) Pick a generating set of S_3 (The permutation group of three elements) and draw its Cayley graph.

Cayley graphs are vertex transitive, but there are vertex transitive graphs that are not Cayley graphs. Here is a an example:

Example 1.78. Consider the graph (called grand-father graph) obtained from the 3-regular tree by choosing a direction orienting the tree and adding the edges linking grand fathers to grand sons (see Fig. 1.2). This graph is vertex-transitive but not a Cayley graph.

Exercise 1.79. (Level 3) Prove the last example is indeed vertex transitive but not a Cayley graph.

It is believed that $p_c < 1$ for all vertex transitive graphs of super linear growth. This is known for all Cayley graphs with exponential growth and also for all known Cayley graphs with intermediate growth (i.e. larger than linear but smaller than exponential). However there is no general argument for Cayley graphs of intermediate growth and the proofs are specific for the known ones. In fact the only Cayley graphs with intermediate growth known are the Grigorchuk groups and the proof that $p_c < 1$ for those groups uses the fact that the two dimensional grid embeds rough isometrically in them, see [MP01].

Agelos Georgakopoulos asked:

Open problem 1.80. *Is it true that one can embed in a rough isometric sense either the two dimensional grid or the binary tree in any superlinear Cayley graph?*

If the answer to the last question is true it implies that $p_c < 1$ for any Cayley graph with superlinear growth.

Next we turn to discuss a very important example of a Cayley graph known as the lamplighter graph.

Example 1.81 (Lamplighter). Imagine there is some person standing on $\mathbb{Z}$ and that at each point $v \in \mathbb{Z}$ there is a lamp that can be either on or off. At each step the person can do one of the following three actions:

1. move one step to the right.
2. move one step to the left.
3. light a lamp in his current location if it was turned off or turn it off if it was turned on.

We start in the position where all lamps are turned off and the person is in the origin. The state space of such a process consists of a vector $\{0, 1\}^{\mathbb{Z}}$ with a finite number of 1's plus an integer representing the current location of the person. The lamplighter graph on $\mathbb{Z}$, denoted by $LL(\mathbb{Z})$, is the graph whose vertices are all possible states of the last process that can be reached in finite number of steps and edges connect two vertices if one can move from one to the other by one step of the process. More precisely: The vertices $V_{LL(\mathbb{Z})}$ is the set of all (w, n) such that $w \in \{\{0, 1\}^{\mathbb{Z}}$ with finite number of 1's$\}$ and $n \in \mathbb{Z}$. For two nodes, $v_1, v_2 \in V_{LL(\mathbb{Z})}$ such that $v_1 = (w_1, k)$ and $v_2 = (w_2, l)$, the edge $\{v_1, v_2\}$ belongs to $E_{LL(\mathbb{Z})}$ if and only one of the following holds:

1. $w_1 = w_2$ and $k = l + 1$.
2. $w_1 = w_2$ and $k = l - 1$.
3. $k = l$, $w_1(j) = w_2(j)$ for all $j \neq k$ and $w_1(k) = 1 - w_2(k)$.

One can show that $LL(\mathbb{Z})$ is a Cayley graph.

In a similar way, one can define the lamplighter graph $LL(G)$ for any graph G.

Example 1.82. The lamplighter graph generated from the circle with n vertices is a cube-connected cycles graph, which is the graph formed by replacing each vertex of the hypercube graph by a cycle, see Fig.,1.3 for the case $n = 3$. The generated graph is a Cayley graph.

The same idea will show that if G is a Cayley graph then $LL(G)$ is also a Cayley graph. In fact there are cases where $LL(G)$ is a Cayley graph though G isn't.

Fig. 1.3 The lamplighter
graph generated from C_3

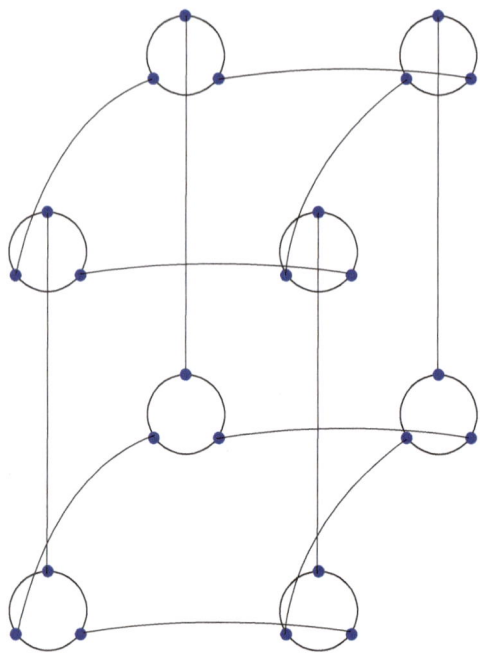

Remark 1.83. It is also possible to consider lamplighter graph with lamps that belong to a larger group than $\mathbb{Z}/2\mathbb{Z}$.

Definition 1.84. The speed exponent of a random walk on a graph is said to be α if $\mathbb{E}[dist(X_n, X_0)] \sim n^\alpha$, up to smaller order terms.

Exercise 1.85. (Level 2) What is the speed of simple random walk on $LL(\mathbb{Z})$ and $LL(\mathbb{Z}^2)$?

Recall Hopf Rinow theorem which state that in any connected Riemannian manifold with complete metric any geodesic can be extended. This is not the case for the lamplighter graph, namely, there are some geodesic intervals that can not be extended.

Exercise 1.86. (Level 2) Describe such a maximal geodesic interval. Try to describe a rooted geodesic tree, consisting of all geodesics from a fixed root.

Remark 1.87. The last fact might help explain the seemingly paradoxical situation of a random walk escaping to infinity diffusively on an exponential volume growth group.

We end this section with some additional remarks and examples on Cayley graphs:

In [Ers03] Anna Erschler initiated the study of speed exponents on groups (for more details see also [LP09]). She proved that the speed exponent of any group is

Fig. 1.4 Petersen graph

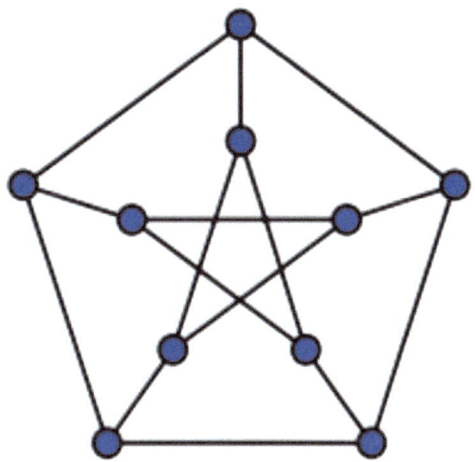

between $\frac{1}{2}$ and 1. She also observed that iterations of the lamplighter groups with $\mathbb{Z}$ lamps gives exponents of the form $1-2^{-k}$ for every $k \in \mathbb{N}$. In a paper in preparation, Amir and Virag construct a group with speed exponent α for every choice of real α between $3/4$ and 1. Constructing groups with speed exponent between $1/2$ and $3/4$ is still open. Calculating the separation profile of lamplighters is also an open question, see [BST12].

Example 1.88 (Heisenberg). Consider 3×3 matrices with integer values, where the diagonal entries are all 1's and below it there are only 0's. This is the smallest noncommutative infinite group, generated by two elements. It can be shown that this group has volume growth with exponent 4, see [GPKS06, Kle10] and references there.

Exercise 1.89 (Petersen Graph). (Level 2) Show that the Petersen graph, see Fig. 1.4, is the smallest vertex transitive graph which is not a Cayley graph.

Example 1.90. The Diestel-Leader graph $D(q, r)$ (see e.g. [Woe05]) is rough isometric to a Cayley graph if and only if $q = r$ [EFW07] (in which case it is a Cayley graph of the lamplighter graph form). Consider one infinite d-ary tree growing downwards, and another r-ary tree growing upwards. A vertex of $D(d, r)$ is a pair of one vertex from each tree, at the same height. (u, v) is connected to (x, y) if $u \sim x$ and $v \sim y$. See Fig. 1.5.

Example 1.91 (The Long Range Graph). In the long range graph the vertices are the integers $\mathbb{Z}$ and the set of edges is $E = \bigcup_{k \geq 0} E_k$ where $E_0 = \{\{i, i+1\}, i \in \mathbb{Z}\}$ and $E_k = \{\{2^k(n-1/2), 2^k(n+1/2)\} : n \in \mathbb{Z}\}$. This graph has subexponential super polynomial growth, see [BH05].

Exercise 1.92. (Level 3) Is the long range graph recurrent? What is its critical percolation parameter?

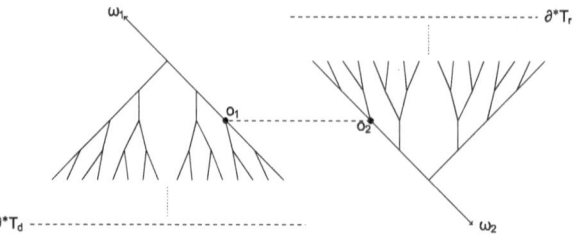

Fig. 1.5 The Diestel-Leader graph. A vertex is a pair (o_1, o_2) of a vertex from each tree

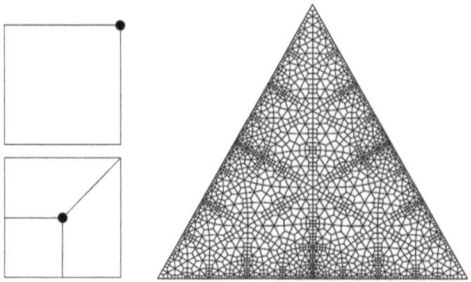

Fig. 1.6 Is there a two sided infinite geodesic in this graph?

Example 1.93 (A Cayley graph with linear growth and large girth). Let G be some finite group with large girth and two generators. Let $\phi_1 : F_2 \to G$ be a quotient map. Let I_1 be its kernel, $I_1 = \phi^{-1}(e)$. Let $\phi_2 : F_2 \to \mathbb{Z}$ be some quotient map, and let I_2 be it's kernel. The group we seek is $F_2/(I_1 \cap I_2)$. It has linear growth since it is finite over $F_2/I_2 = \mathbb{Z}$. It has high girth since taking smaller divisor only increases the girth.

Example 1.94 (Recursive Subdivisions). It is possible to construct planar triangulations with uniform growth r^α for every $\alpha > 1$. Here is an example of a quadrangulation of the plane obtained by starting with a quadrilateral with a marked corner, subdividing it as in Fig. 1.6 to obtain three quadrilaterals with the interior vertex as the marked corner of each, and continuing inductively. The result is a map of the plane with quadrilateral faces and maximum degree 6.

Chapter 2
On the Structure of Vertex Transitive Graphs

This short section contains several facts and open problems regarding vertex transitive graphs, starting with the following theorem from [BS92] which refines an earlier result of Aldous.

Recall a graph is *vertex transitive* if for any pair of vertices in the graph, there is a graph automorphism mapping one to the other.

Theorem 2.1. *Let $G = (V, E)$ be a finite vertex transitive graph with diameter d. Then for any subset $S \subset V$*

$$\mu(\partial S) \geq \frac{2\mu(S)\mu(S^c)}{d + \mu(S)}$$

where S^c is $V \backslash S$, ∂S is the outer vertex boundary of S and $\mu(S) = \frac{|S|}{|V|}$.

Proof. Choose a random (ordered) pair of vertices (x, y) (uniformly from $V \times V$) and randomly choose a shortest path γ between them (again uniformly from the set of all such paths). From the vertex-transitivity of G we infer that the probability that γ goes through a given vertex $z \in V$ is $\frac{D+1}{|V|}$, where D is the expected distance between a random pair of vertices and $|V|$ is the number of vertices in G. Since a path between a vertex from S and a vertex from S^c must intersect the boundary of S at least ones, it follows that

$$\frac{|\partial S|(D+1)}{|V|} = \mathbb{P}(\gamma \text{ goes through } \partial S) \qquad (2.1)$$
$$\geq \mathbb{P}(x \in S, y \in S^c \text{ or } x \in S^c, y \in S \cup \partial S).$$

The r.h.s. equals $(2\mu(S) + \mu(\partial S))\mu(S^c)$, while the l.h.s. is bounded from above by $(d + 1)\mu(\partial S)$, since $d \geq D$. Thus the theorem follows from the fact that $\mu(S) + \mu(S^c) = 1$. □

I. Benjamini, *Coarse 3Geometry and Randomness*, Lecture Notes in Mathematics 2100,
DOI 10.1007/978-3-319-02576-6_2,
© Springer International Publishing Switzerland 2013

As a corollary we get that for any $|S| \leq \frac{|V|}{2}$,

$$\frac{|\partial S|}{|S|} \geq \frac{2}{2d+1}.$$

This is in contrast with some natural random graphs such as the Uniform infinite planar triangulation and long range percolation, that can admits bigger bottlenecks.

2.1 Scaling Limits

Recall that the *Gromov-Hausdorff distance* between two metric spaces is obtained by taking the infimum over all the Hausdorff distances between isometric embeddings of the two spaces in a common metric space.

Is there a sequence of finite vertex transitive graphs that converge in the Gromov-Hausdorff metric to the sphere S^2? (equipped with some invariant proper length metric).

Let $\{G_n\}$ be sequence of finite, connected, vertex transitive graphs with bounded degree such that $|G_n| = o(\operatorname{diam}(G_n)^d)$ for some $d > 0$.

With Hilary Finucane and Romain Tessera, see [BFT12], we proved,

Theorem 2.2. *Up to taking a subsequence, and after rescaling by the diameter, the sequence $\{G_n\}$ converges in the Gromov Hausdorff distance to a torus of dimension $< d$, equipped with some invariant proper length metric.*

In particular if the sequence admits a doubling property at a small scale then the limit will be a torus equipped with some invariant proper length metric. Otherwise it will not converge to a finite dimensional manifold.

The proof relies on a recent quantitative version of Gromov's theorem on groups with polynomial growth obtained by Breuillard et al. [BGT12] and a scaling limit theorem for nilpotent groups by Pansu.

A quantitative version of this theorem can be useful in establishing the resistance conjecture from [BK05] and the polynomial case of the conjectures in [ABS04]. In [BGT12] a strong isoperimetric inequality for finite vertex transitive is established.

If $\{G_n\}$ are only roughly transitive and $|G_n| = o(\operatorname{diam}(G_n)^{1+\delta})$ for $\delta > 0$ sufficiently small, we are able to prove, this time by elementary means, that $\{G_n\}$ converges to a circle.

2.2 Questions

Open problem 2.3. *A metric space X is C-roughly transitive if for every pair of points $x, y \in X$ there is a C-rough-isometry sending x to y. Is there an infinite C-roughly transitive graph, with C finite, which is not roughly-isometric*

to a homogeneous space, where a homogeneous space is a space with a transitive isometry group.

The following two questions are regarding the rigidity of the global structure given local information.

Open problem 2.4 (With Agelos Georgakopoulos). *Is it the case that for every Cayley graph G there is $r = r(G)$ such that G covers every r-locally-G graph? Here we say that a graph H is r-locally-G if every ball of radius r in H is isomorphic to the ball of radius r in G.*

Open problem 2.5. *Given a fixed rooted ball $B(o, r)$, assume there is a finite graph such that all its r-balls are isomorphic to $B(o, r)$), e.g. $B(o, r)$ is a ball in a finite vertex transitive graph, what is the minimal diameter of a graph with all of its r-balls isomorphic to $B(o, r)$? Any bounds on this minimal diameter, assuming the degree of o is d? Any example where it grows faster than linear in r, when d is fixed?*

When the rooted ball is a tree, this is the girth problem.

Exercise 2.6. (Level 4) Show that for some r, the r-ball in the grandparent graph Example 1.78, does not appear as a ball in a finite vertex transitive graph.

Not assuming a bound on the degree, consider the 3-ball in the hypercube, is there a graph with a smaller diameter than the hypercube so that all its 3-balls are that of the hypercube?

Here are some additional open problems on transitive graphs:

Open problem 2.7.

1. *Let G be a finite graph of size n. Given $k < n$, observe the set of balls in G of size k. Assume that as rooted graphs all these balls are isometric, Does it imply that G is vertex transitive?*
2. *Here is an example that shows that this is false if we choose k small enough: Assume G is a union of two cycles one of size $\frac{n}{2} + 1$ and the other of size $\frac{n}{2} - 1$. This imply that all balls of size $\frac{n}{2} - 3$ are rooted intervals (when considered as rooted graphs), and in particular isomorphic. On the other hand it is easy to see that G is not vertex transitive.*
3. *What if k is larger than $\frac{n}{2}$? Say $0.99n$, or the diameter of the graph minus some constant.*
4. *Let n be odd and look at the family of all vertex transitive graphs with n vertices. Since the complement, i.e. the graph with the same set of vertices and the complement of the set of edges with respect to the full set of edges, of any such graph is also vertex transitive, it follows that the expected degree of random uniformly vertex transitive graph is $\frac{n-1}{2}$. We conjecture that it is concentrated near $\frac{n-1}{2}$.*

Chapter 3
The Hyperbolic Plane and Hyperbolic Graphs

The aim of this section is to give a very short introduction to planar hyperbolic geometry. Some good references for parts of this section are [CFKP97] and [ABC+91]. We first discuss the hyperbolic plane. Nets in the hyperbolic plane are concrete examples of the more general hyperbolic graphs. Hyperbolicity is reflected in the behaviour of random walks [Anc88] and percolation as we will see in Chap. 7.

To get an *intuitive feel* for the hyperbolic plane, consider the graph obtained by adding edges to a d-regular tree, $(d > 2)$, creating a cycle between all the vertices with the same distance to a fixed root. This graph is rough isometric to the hyperbolic plane.

Exercise 3.1. (Level 3) Prove this.

Exercise 3.2. (Level 3) What is the separation function (in the sense of Sect. 1.4) of this graph?

3.1 The Hyperbolic Plane

There are several models of hyperbolic geometry. All of them are equivalent, in the sense that there are isometries between them. Which model one wants to work with very much depends on the nature of the problem of interest. The most common models are the Poincaré unit disc model and the half plane model. We now concentrate on the properties of the first.

Denote a point in the complex plane $\mathbb{C}$ by $z = x + iy$. The Poincaré disc model of hyperbolic space is given by the open unit disc $\mathbb{D} = \{z \in \mathbb{C} : |z| < 1\}$ equipped with the metric, which we refer to as the hyperbolic metric,

$$ds^2 := 4\frac{dx^2 + dy^2}{(1 - x^2 - y^2)^2}. \tag{3.1}$$

I. Benjamini, *Coarse 3Geometry and Randomness*, Lecture Notes in Mathematics 2100,   23
DOI 10.1007/978-3-319-02576-6_3,
© Springer International Publishing Switzerland 2013

We denote this space by $\mathbb{H}^2$, sometimes called the hyperbolic plane. From (3.1) we see that near the origin, ds^2 behaves like a scaled Euclidean metric, but there is heavy distortion near the boundary of $\mathbb{D}$. The factor 4 in (3.1) is often omitted from the definition of the hyperbolic metric. We remark that it is also common to identify points of $\mathbb{H}^2$ with points in the open unit disc in the Euclidean plane rather than in the complex plane.

In the hyperbolic metric, a curve $\{\gamma(t) : 0 \leq t \leq 1\}$ has length

$$L(\gamma) = 2 \int_0^1 \frac{|\gamma'(t)|}{1 - |\gamma(t)|^2} dt$$

and a set A has area

$$\mu(A) = 4 \int_A \frac{dx\,dy}{(1 - x^2 y^2)^2}.$$

If $z_1, z_2 \in \mathbb{H}^2$, then the geodesic between them (that is, the shortest curve that starts at z_1 and ends at z_2) is either a segment of an Euclidean circle that intersects the boundary of $\mathbb{D}$ orthogonally, or a segment of a straight line that passes through the origin. Recall that Euclid's parallel postulate says that given a line and a point not on it, there is exactly one line going through the given point that is parallel to the given line. The space $\mathbb{H}^2$ does not satisfy Euclid's parallel postulate which means $\mathbb{H}^2$ has a non-Euclidean geometry.

Let us consider some areas and lengths in this metric. Let $z_1, z_2 \in \mathbb{H}^2$. The hyperbolic distance between z_1 and z_2 is given by

$$d(z_1, z_2) = 2 \tanh^{-1} \left(\left| \frac{z_2 - z_1}{1 - \bar{z}_1 z_2} \right| \right).$$

Let $B(x, r) := \{y \in \mathbb{H}^2 : d(x, y) \leq r\}$ be the closed hyperbolic ball of radius r centered at x. The circumference of the ball is given by

$$L(\partial B(x, r)) = 2\pi \sinh(r)$$

and the area is given by

$$\mu(B(x, r)) = 2\pi(\cosh(r) - 1). \tag{3.2}$$

Observe that

$$2\pi \sinh(r) = 2\pi r + o(r^2) \text{ as } r \to 0 \tag{3.3}$$

and

$$2\pi(\cosh(r) - 1) = \pi r^2 + o(r^3) \text{ as } r \to 0. \tag{3.4}$$

This implies that the formulas are well approximated with the Euclidean formulas at a small scale. Also, we see that as $r \to \infty$, both the area and circumference grow exponentially with the same rate. Moreover, the ratio between them tends to 1 as $r \to \infty$. In fact, if A is any bounded set for which $\mu(A)$ and $L(\partial A)$ are well defined, we have

$$L(\partial A) \geq \mu(A). \tag{3.5}$$

This is the so called linear isoperimetric inequality for $\mathbb{H}^2$. Such an inequality is not available in the Euclidean plane.

Next, we consider tilling of $\mathbb{H}^2$. Recall that two sets are said to be congruent if there is an isometry between them. A regular tiling of a space is a collection of congruent polygons that fill the space and overlap only on a set of measure 0, such that the number of polygons that meet at a corner is the same for every corner. For example, there are exactly three kinds of such tilings for the Euclidean plane. These are made up of equilateral triangles, squares or hexagons (however, given any side length, a regular tiling of any of these types exist). In the hyperbolic plane the situation is different. There exists an infinite number of regular tilings. More precisely, if p and q are positive integers such that $(p - 2)(q - 2) > 4$, then it is possible to construct a regular tiling of the hyperbolic plane with congruent p-gons, where at each corner exactly q of these p-gons meet. However, given p and q, there is only one choice for the side length of the p-gon that gives this tiling. Each regular tiling of $\mathbb{H}^2$ can be identified with a graph G. More precisely, each side in the tiling is identified with an edge in G and each corner is identified with a vertex in G. Such a graph is transitive, and has a positive Cheeger constant.

One more useful fact about the Poincaré disc model is the following: Consider the ball $B(x, r)$ in $\mathbb{H}^2$. This ball actually looks precisely like an Euclidean ball. However, its Euclidean center is closer to origin than its hyperbolic center x, and its Euclidean radius is smaller than its hyperbolic radius r. There are explicit formulas for both these quantities.

A comment regarding the Poincaré half plane model. This model for the hyperbolic plane is given by the complex upper half plane $\{x + iy : y > 0\}$ together with the metric

$$ds^2 := 4\frac{dx^2 + dy^2}{y^2}.$$

In this model, the intersection of the upper half plane with Euclidean circles orthogonal to the real line are infinite geodesics. An isometry from the half plane model to the unit disc model is given by $f(z) = \frac{z-i}{z+i}$, and the inverse of this isometry is given by $f^{-1}(z) = i\frac{1+z}{1-z}$.

Hyperbolic space is an example of a symmetric space. A symmetric space is a connected Riemannian manifold M, such that for every point $p \in M$, there is an isometry I_p such that $I_p(p) = p$ and I_p reverses all geodesics through p. In $\mathbb{H}^2$, such an isometry is simply given by a rotation of $180°$, around the point p.

Other symmetric spaces are the Euclidean space and the sphere (in any dimensions). Symmetric spaces belong to the class of homogeneous spaces.

3.2   Hyperbolic Graphs

We now move to the wider set up of hyperbolic graphs, which form a large class of hyperbolic spaces. Let us start by defining hyperbolic spaces and state some of their basic properties. The most general definition uses the notion of the Gromov product.

Definition 3.3. Let (X, d) be a metric space and $x, y, z \in X$ three points in it. The Gromov product $(x|y)_z$ of x and y with respect to z is defined by

$$(x|y)_z = \frac{1}{2} \left(d(x, z) + d(y, z) - d(x, y) \right).$$

The geometric intuition behind $(x|y)_z$ is the following: Up to some additive constant it describes the distance from z to any x-y geodesic. A good exercise for the reader at this point is to check that if X is a tree then $(x|y)_z$ is in fact precisely this distance.

We now give the definition of δ-hyperbolic space.

Definition 3.4. A metric space (X, d) is called δ-hyperbolic if for every four points $x, y, z, w \in X$ the following inequality holds

$$(x|z)_w \geq \min\{(x|y)_w, (y|z)_w\} - \delta.$$

This definition can be rewritten in another form. There exist three possibilities to divide these four points into pairs. Consider the corresponding sums of distances

$$p = d(x, w) + d(z, y) \quad , \quad m = d(x, y) + d(z, w) \quad , \quad g = d(x, z) + d(y, w).$$

Rename the points if needed to ensure that $p \leq m \leq g$. Then Definition 3.4 can be rewritten in the following form

$$g \leq m + 2\delta.$$

In other words, the greatest sum cannot exceed the mean sum by more than 2δ.

If our space (X, d) is geodesic we can use one more equivalent definition for δ-hyperbolicity, this time in terms of "thin triangles". For two given points $x, y \in X$ we will denote by xy a geodesic segment between them. In general such a geodesic segment is not necessarily unique so under this notation we assume xy is one of these geodesic segments.

Definition 3.5. A geodesic triangle xyz formed by xy, yz and zx is called δ-thin if the distance from any point p in xy to the union of xz and yz does not exceed δ, i.e.

$$\sup_{p \in xy} d(p, xz \cup yz) \le \delta.$$

Proposition 3.6. *A geodesic metric space* (X, d) *is* δ-*hyperbolic if and only if every geodesic triangle is* $\frac{1}{2}\delta$-*thin.*

According to Bonk and Schramm [BS00], every δ-hyperbolic metric space can be embedded isometrically into a complete δ-hyperbolic geodesic metric space. So, many theorems can be reduced to the investigation of geodesic hyperbolic spaces using the definition of hyperbolicity in terms of δ-thin triangles.

Next we introduce the notion of a *divergence function* which allows to estimate lengths of paths connecting two diverging geodesics outside a ball as a function of the radius of that ball. Later this approach will help us to show that the length of a curve lying far away from a geodesic is very big.

Definition 3.7. Let (X, d) be a metric space. We say that $\eta : \mathbb{N} \to \mathbb{R}$ is a *divergence function* for the space (X, d) if for any point $x \in X$ and any two geodesic segments $\gamma = xy$ and $\gamma' = xz$ the following holds: For any $r, R \in \mathbb{N}$ such that the lengths of γ and γ' exceed $R + r$, if $d(\gamma(R), \gamma'(R)) > \eta(0)$ and σ is a path from $\gamma(R + r)$ to $\gamma'(R + r)$ in the closure of the complement of the ball $B_{R+r}(x)$ (that is in $X \setminus B_{R+r}(x)$), then the length of σ is greater than $\eta(r)$.

In any metric space, when two points move along two geodesic rays, the distance between them grows linearly by the triangle inequality. However, if instead of the distance between two such points x_n, y_n we consider the length of the shortest x_n, y_n path outside a ball of radius n around their common origin, it turns out that the lengths of these paths grow exponentially if our space is hyperbolic (for example the length of a circle grows exponentially with the radius). If our space (X, d) admits an exponential diverge function then we say that geodesics diverge exponentially in (X, d).

Theorem 3.8. *In a hyperbolic space geodesics diverge exponentially.*

An amazing fact is that the opposite statement is also true and even more: a non-linear divergence function in a geodesic space implies the existence of an exponential divergence function, and so, the space is hyperbolic. We are not going to prove this theorem here.

Proof. As in Definition 3.7 let γ and γ' be two geodesics of length $R + r$ with one end at the same point x and such that $d(\gamma(R), \gamma'(R)) > 4\delta$. We assume $\eta(0) = \delta$. Let σ be a path connecting the ends $\gamma(R + r)$ and $\gamma'(R + r)$ which lie outside of $B_{R+r}(x)$. We have to show that there exists an exponential function $\eta(r)$ independent of γ and γ' such that $len(\sigma) \ge \eta(r)$.

Let α be a geodesic connecting $\gamma(R + r)$ and $\gamma'(R + r)$. For the following constriction we will use binary sequences b which are sequences of 1 and 0 of length

s (the zero-length sequence is also allowed, i.e., $b = \emptyset$). For every binary sequence b we define a geodesic which we denote by α_b. First $\alpha_\emptyset = \alpha$. Next assume that we have already constructed the geodesics α_b for every b of length not exceeding s (it will follow from the construction that the ends of these geodesics lie on the curve σ). For every b of length not more then s denote the midpoint of the segment of σ between the ends of α_b by m_b. We define α_{b0} to be a geodesic connecting m_b with $\alpha_b(0)$ and α_{b1} to be a geodesic connecting m_b with $\alpha_b(1)$.

We continue this process until we obtain a subdivision consisting of α_b with lengths between $\frac{1}{2}$ and 1. Such a subdivision will be obtained after at most $\log_2(len(\sigma)) + 1$ steps. Recall that since the space is hyperbolic, all the triangles with the sides α_b, α_{b0}, α_{b1} are δ-thin. Since $d(\gamma(R), \gamma'(R)) > 4\delta$ it follows that $d(\gamma(R), \gamma) > \delta$ and hence there exists a point $v(0)$ on α such that $d(\gamma(R), v(0)) < \delta$. Now either on α_0 or on α_1 we can find a point $v(1)$ at distance less than δ from $v(0)$. We Continue in the same manner. If we constructed a sequence of points $\{v(i)\}$ with $0 \le i \le n$ with $n < \log_2(len(\sigma)) + 1$ and $v(n)$ lies on some geodesic of length not greater than 1. Then we can find a point $v(n + 1)$ on σ with $d(v(n), v(n + 1)) < 1$. We can estimate the distance from x to $v(n + 1)$ by the "length" of the chain $v(i)$

$$d(x, v(n + 1)) \le R + \delta(\log_2(len(p)) + 1) + 1.$$

On the other hand $d(x, v(n + 1)) > R + r$ by definition. Hence,

$$r \le \delta(\log_2(len(p)) + 1) + 1,$$

which implies

$$len(p) > 2^{\frac{r-1}{\delta} - 1}$$

and completes the proof.   □

A metric tree is one of the most basic examples of a hyperbolic space. Most of the properties of hyperbolic spaces can be illustrated in trees and theorems in this subject should be first verified for them. The following theorem establishes the close relation between hyperbolic spaces and trees. It says that if we are looking from far away then a hyperbolic space looks similar to a tree. Given a set A in a metric space (X, d) we denote by diam(A) for the diameter of A.

Theorem 3.9. *Let (X, d) be a metric δ-hyperbolic space with a base point w and let k be a positive integer. If diam$(X) \le 2^k + 2$ then there exist a finite metric tree (T, d_T) with a base point t and a map $\Phi : X \to T$ such that*

1. Φ preserves distances to the base point, i.e.

$$d_T(\Phi(x), t) = d(x, w),$$

for any point x of X.

2. *For any two points* $x, y \in X$

$$d(x, y) - 2k\delta \leq d_T(\Phi(x), \Phi(y)) \leq d(x, y).$$

Proof. We will give the main ideas of the proof here, for the full details see [GH90] Chap. 2 (you will also find a more general variant of this theorem which admits existence of rays in this space). Consider an integer L and a sequence $x_1, \ldots, x_L$ of points in X. By mathematical induction (divide the chain into two chains of lengths not exceeding $2^{k-1} + 1$) we prove that

$$(x_1|x_L) \geq \min_{2 \leq i \leq L} (x_{i-1}|x_i) - k\delta.$$

Now we will define a new pseudometric on X which is 0-hyperbolic. First introduce the following notation

$$(x|y)' = \sup \left\{ \min_{2 \leq i \leq L} (x_{i-1}|x_i) \right\},$$

we take the supremum by all chains connecting $x = x_1$ and $y = x_L$. The new pseudometric (verify that it is really a pseudometric) is

$$|x - y|' = |x - w| + |y - w| - 2(x|y)'.$$

Denote also $(k + 1)\delta - 2c$ by δ'. This pseudometric

- is 0-hyperbolic, that is $(x|z)' \geq \min \{(x|y)', (y|z)'\}$ for every $x, y, z \in X$;
- is in a bounded distance from the initial metric:

$$|x - y| - 2\delta' \leq |x - y|' \leq |x - y|;$$

- preserves the initial distances to the base point:

$$|x|' = |x|.$$

Now consider the quotient space F' of F by the equivalence relation $\sim$:

$$x \sim y \Leftrightarrow |x - y|' = 0.$$

Hence $| \cdot |'$ defines a metric on F'. It is known (see for example [GH90] Chap. 2, Proposition 6) that every finite 0-hyperbolic space can be isometrically embedded in a metric tree T. So the composition of natural maps $F \to F'$ and $F' \to T$ satisfies the conditions of the theorem.   □

We now introduce a class of maps which will allow us to compare hyperbolic spaces.

Definition 3.10. Two metric spaces (X, d_X) and (Y, d_Y) are said to be rough isometric if there are two maps $f : X \to Y$, $g : Y \to X$ and two constants $\lambda > 0$ and $c \geq 0$ such that

1. $|f(x) - f(y)| \leq \lambda |x - y| + c$ for every $x, y \in X$,
2. $|g(x') - g(y')| \leq \lambda |x' - y'| + c$ for every $x', y' \in Y$,
3. $|g(f(x)) - x| \leq c$ for every $x \in X$,
4. $|f(g(x')) - x'| \leq c$ for every $x' \in Y$.

The first two conditions mean that f and g are nearly Lipschitz if we are looking from far away. The two other conditions mean that f and g are nearly inverse of each other. It is easy to check that the composition of two rough-isometries is also a rough-isometry. Thus, rough isometries provide an equivalence relation on the class of metric spaces.

Exercise 3.11. (Level 2)

1. Compare the above definition with Theorem 3.9. Find rough isometry constants between X and T.
2. Show that a regular (infinite) tree and a hyperbolic plane are not rough isometric.

Consider a finitely generated group G with a symmetric generating set S. Introduce the following metric d_S (which is called a *word metric*) on G:

$$d_S(g_1, g_2) = \min \left\{ k \ : \ g_1^{-1} g_2 = s_1 s_2 \ldots s_k, \text{ such that } s_i \in S, \forall 1 \leq i \leq k \right\}.$$

In other wordsj, $d_S(g_1, g_2)$ is the graph-distance between g_1, g_2 in the Cayley graph of G with respect to S. The *length* of an element $g \in G$ is its distance from e, that is the minimal number of generators which are needed to represent g.

Exercise 3.12. (Level 1) The word metric is a left-invariant metric. Start by proving that it is well-defined.

A finitely generated group G provides an important example of rough isometric spaces. Considering two finite generating sets S_1, S_2, we obtain two different metric spaces $< G, S_1 >$ and $< G, S_2 >$ which are rough isometric to each other. Indeed, let λ_1 be the maximal value of $d_{S'}$ on the set S and λ_2 the maximal value of d_S on the set S'. Then,

$$d_{S'} \leq \lambda_1 d_S, \quad \text{and} \quad d_S \leq \lambda_2 d_{S'}.$$

Definition 3.13. A map $f : (E, d_E) \to (F, d_F)$ between metric spaces is a *rough* (λ, c)-*rough-isometric embedding*, if for any two points $x, y \in E$ we have

$$\frac{1}{\lambda} d_E(x, y) - c \leq d_F(f(x), f(y)) \leq \lambda d_E(x, y) + c.$$

This definition follows from the definition for two spaces being rough isometric but it does not include the existence of a nearly inverse map. We can easily transform Definition 3.13 to make it equivalent to Definition 3.10 by adding the condition that f is nearly surjective, i.e., for every point $y \in F$ there exists a point $x \in E$ such that $d_F(y, f(x)) < c$.

Definition 3.14. A (λ, c)-quasi-geodesic in F is a (λ, c)-rough isometry from a real interval $I = [0, l]$ to F.

One of the most important theorems characterizing quasi-geodesics in hyperbolic spaces is Morse Lemma which says that a quasi-geodesic lies near a geodesic connecting its ends. In addition the distance between them depends only on the constants of the quasi-isometry λ and c and on the hyperbolic constant δ.

Theorem 3.15. *Let F be a δ-hyperbolic space, γ a (λ, c)-quasi-geodesic and σ a geodesic connecting the ends of γ. Then γ is in a H-neighbourhood of σ where $H = H(\lambda, c, \delta)$.*

Proof. It is possible to prove, see [Shc13], that for any (λ, c)-quasi-geodesic γ there exists a continuous (λ, c')-quasi-geodesic γ' lying in c_1-neighbourhood of γ where c', c_1 are bounded by several times c. Moreover, the length of any arc L of such a quasi-geodesic is bounded by $4\lambda^2 R$ where R is the distance between the ends of this arc. So, in this proof we will assume that γ is a continuous quasi-geodesic.

First we will prove that the geodesic σ lies in H_1-neighbourhood of γ where H_1 depends only on δ. Assume that z is the point in σ which is most distant from γ and that its distance from γ is L. Denote by a and b the points on σ at distance L from z, one in each direction of σ (if the length of σ is insufficient then just take its ends). In the same manner denote by a_1, b_1 the points on σ at distance $2L$ from z. Find the closest to a_1, b_1 points of γ, denote them by a', b'. From the definition of L and the points we have $d(a_1, a'), d(b_1, b') \leq L$. Moreover, the geodesic segments $a_1 a', b_1 b'$ lie in the complement of the ball $B_L(z)$. Hence the path which consists of $a_1 a'$, the part of the quasi-geodesic γ between a', b' and $b' b_1$ is a path which lies in the complement of the ball $B_L(z)$. Thus its length should be greater than e^L. On the other hand as we have mentioned in the beginning of the proof the length of this path is less than $2L + 16L\lambda^2$. Combining these inequalities we conclude that L is bounded by some $H_1 = H_1(\delta, \lambda, c)$.

Now assume that $a' b'$ is a part of γ lying in the complement of the H_1-neighbourhood of σ. Denote the ends of γ (which are the ends of σ at the same time) by p_1, p_2. By the first part of the proof we conclude that σ is in the H_1-neighbourhood of the union of two parts of γ: $(p_1 a')$ and $(b' p_2)$. Hence there exists such a point p of σ that $d(p, (p_1 a')), d(p, (p_2 b')) < L$. Hence, the length of the part $(a' b')$ cannot exceed $2H_1\lambda^2$. Finally we obtain that there exists an upper bound $H = L + 2H_1\lambda^2$ for the distance from any point of the quasi-geodesic γ to the geodesic σ what finishes the proof.   □

Exercise 3.16. (Level 4) Prove that the δ-hyperbolicity constant of finite vertex transitive graphs and expanders, is proportional to it's diameter.

Chapter 4
Percolation on Graphs

In this section we introduce and discuss some basic properties of percolation, a fundamental random process on graphs. For background on percolation see [Gri99].

4.1 Bernoulli Percolation

A percolation process is a random diluting of a graph. There are various versions of percolation processes. For sake of simplicity, in this section will we focus on the simplest one, the so-called Bernoulli (or independent) bond percolation. More general percolation processes are considered later, see Definition 8.9.

Let $G = (V, E)$ be a graph (finite or infinite), and fix a parameter $p \in [0, 1]$. We associate a Bernoulli random variable X_e with every edge $e \in E$, given by

$$X_e = \begin{cases} 1 & \text{with probability } p \\ 0 & \text{with probability } 1 - p \end{cases},$$

such that the $\{X_e\}_{e \in E}$ are i.i.d. By a percolation process we mean the diluted graph generated by removing all edges $e \in E$ such that $X_e = 0$. The remaining subgraph might not be connected and its connected components are called open clusters (or open components). We will denote the probability measure associated with this process by $\mathbb{P}_p$. Below we will often talk about components, or more precisely open components, obtained by this process. The component of a vertex v is defined as the set of all vertices that can be reached from v using a path of open edges, i.e. the set of vertices $w \in V$ such that $\exists n \in \mathbb{N}$ and $v_0, v_1, \ldots, v_n \in V$ satisfying $v_0 = v$, $v_n = w$, $\{v_i \, v_{i+1}\} \in E$ and $X_{v_i.v_{i+1}} = 1$ for every $0 \leq i \leq n - 1$. We often regard edges for which $X_e = 1$ as open, and edges for which $X_e = 0$ as closed.

I. Benjamini, *Coarse 3Geometry and Randomness*, Lecture Notes in Mathematics 2100, DOI 10.1007/978-3-319-02576-6_4, © Springer International Publishing Switzerland 2013

4.2   The Critical Probability

Fix a graph G with a distinguished vertex ρ which we call the root of the graph.
When talking about the graph $\mathbb{Z}^d$ we always choose the root as the origin $0 = 0_{\mathbb{Z}^d}$.
We say that the vertex ρ is connected to ∞ in the percolation process, and denote this
event by $\rho \leftrightarrow \infty$, if it is contained in an infinite open component (formed by open
edges only). In this case we say the percolation occurs or that the graph percolated.
The following theorem is the stepping-stone in the theory of percolation:

Theorem 4.1. *Bernoulli bond percolation on $\mathbb{Z}^d$ satisfies the following:*

1. If $d > 1$ and $p > 3/4$ then 0 is connected to ∞ with positive probability.
2. If $d = 2$ and $p < 1/4$ then 0 is not connected to ∞ with probability 1.

Notice that it is enough to show the first part of the theorem for $d = 2$, since
we can think of $\mathbb{Z}^2$ as a subgraph of $\mathbb{Z}^d$ for $d \geq 3$. The above theorem suggests the
existence of a phase transition concerning the existence of an infinite cluster in the
percolation process.

Definition 4.2. Let G be a graph with a fixed root ρ. We define

$$p_c = p_c(G, \rho) = \inf\{0 \leq p \leq 1 : P_p(\rho \leftrightarrow \infty) > 0\}.$$

This is called the critical probability for Bernoulli bond percolation on G rooted
at ρ.

Exercise 4.3. (Level 1) Show that if G is a connected graph then $p_c(G, \rho)$ does not
depend on the choice of the root $\rho \in G$.

Thanks to the above exercise we can speak about the critical parameter $p_c(G)$ of
a given connected graph. With this new notation, we may rephrase the last theorem
as $1/4 < p_c(\mathbb{Z}^2) < 3/4$, which would also imply that $p_c(\mathbb{Z}^d) < 3/4$ for every
$d > 2$ (since if $H \subset G$ then $p_c(G) \leq p_c(H)$).
    In the following, unless explicitly mentioned, all the graphs considered are
infinite and connected.

Exercise 4.4. (Level 1) Does there exists an infinite connected graph G with
$p_c(G) = 1$? Hint: Consider $\mathbb{Z}$.

Recall that the Cartesian product of two graphs G and H is the graph denoted
$G \times H$ with vertex set $V = \{(u, v) : u \in G, v \in H\}$ and edge set

$$E = \{\{(u, v), (x, y)\} : u = x \text{ and } \{v, y\} \in E_H \text{ or } v = y \text{ and } \{u, x\} \in E_G\}.$$

Exercise 4.5. (Level 2) Calculate $p_c(\mathbb{Z} \times \mathbb{Z}_2)$. (Hint: from far away, this graph looks
similar to $\mathbb{Z}$).

Theorem 4.6. *Let G be a connected graph with* $\deg(v) \leq d$ *for every* $v \in V$. *Then*

$$p_c(G) \geq \frac{1}{d-1}.$$

Proof. Let ρ be a distinguished vertex in G. If $\rho \leftrightarrow \infty$ then there exists an open self avoiding walk (a path using each edge at most once) from ρ to ∞. The number of self avoiding walks (SAW) of length n starting at ρ in a graph whose degrees are bounded by d is at most $d(d-1)^{n-1}$ (in fact, this bound is tight for a d-regular tree). The probability for any such path to be open is p^n. Hence,

$$\mathbb{P}_p(\rho \leftrightarrow \infty) \leq \mathbb{P}_p\big(\text{there exists an open SAW of length } n \text{ from } 0\big) \leq d(d-1)^{n-1}p^n.$$

If $p < \frac{1}{d-1}$ then the last expression tends to 0 as n tends to ∞, which completes the proof. $\square$

Corollary 4.7.

- We have $p_c(d\text{-regular tree}) \geq \dfrac{1}{d-1}$.
- $p_c(\mathbb{Z}^d) \geq \frac{1}{2d-1}$. In particular for $d = 2$ we get $p_c(\mathbb{Z}^2) \geq \frac{1}{3}$, which proves part 2 of Theorem 4.1.

Exercise 4.8. (Level 2) Use branching process theory (in particular extinction criterion for Galton-Watson trees) to show that $p_c(d\text{-regular tree}) = \frac{1}{d-1}$.

Definition 4.9. For a graph G with some root vertex ρ, we say that a set of edges is a *cut set* if it separates the root from ∞, i.e., any path from the root to ∞ must cross and edge from the set. For example, if we take $\mathbb{Z}$ with the root vertex 0, then $\{\{-3,-2\}, \{5,6\}\}$ is a cut set while $\{\{3,4\}, \{10,11\}\}$ is not.

Definition 4.10. A cut set S of a rooted graph (G, ρ) is said to be a Minimal Cut Set (MCS) if it is a cut set and any $T \subsetneq S$ is not a cut set.

Next we present the first sufficient condition for $p_c < 1$ which is based on the notion of minimal cut sets.

Theorem 4.11. *Let G be a connected infinite graph with a root vertex ρ. If there exists a constant $C > 0$ such that for every $n \geq 0$ we have* #{*MCSs of size* n} < C^n, *then* $p_c(G) < 1$.

Proof. If no percolation occurs, that is ρ doesn't belong to an infinite open component, then there exist an MCS all of its edges are closed in the percolation process. If S is a cut set of size n, the probability that it is closed is $(1-p)^n$. Applying a union bound argument and using the assumption we get

$$\mathbb{P}_p(\text{there exists a closed MCS of size } n) \leq (C(1-p))^n.$$

Hence,

$$\mathbb{P}_p(\text{there exists a closed MCS }) \le \sum_{n=1}^{\infty}(C(1-p))^n.$$

Since the last expression is finite for large enough p and goes to 0 as $p \to 1$ we can find a $p_0 < 1$ such that for any $p \ge p_0$, there is an open path to infinity with positive probability.   □

Exercise 4.12. (Level 2) Let G be a graph with bounded degree. Prove that if $h(G) > 0$ then for every root vertex ρ we have #{MCSs of size n} $< C^n$ for some $C > 0$.

In particular, thanks to the last Exercise and Theorem 4.11 we deduce that if G is a bounded degree graph with $h(G) > 0$, then $p_c(G) < 1$. Note that the condition of Theorem 4.11 can also be applied directly even if $h(G) = 0$.

Exercise 4.13. (Level 2) Use Theorem 4.11 to show that $p_c(\mathbb{Z}^2) < 1$.

Conjecture 4.14. If $\mathrm{I} - \dim(G) > 1$ then $p_c(G) < 1$. A weaker conjecture if $\mathrm{I}\text{-}\dim(G) = \infty$ then $p_c(G) < 1$.

Exercise 4.15. (Level 3) Show that the property $p_c < 1$ is invariant under rough isometries between bounded degree graphs, see Chap. 6 for definitions. Hint: Use domination by product measure (see [LSS97]).

4.3 Exponential Intersection Tail

In this section, we discuss another method for proving that $p_c(G) < 1$, based on an idea of Kesten. Much of the material in this section can be found in [BPP98] and [Per99]. Let γ_1, γ_2 be two paths of Self-Avoiding Walks (SAW), and denote by $|\gamma_1 \cap \gamma_2|$ the number of edges in their intersection.

Definition 4.16. A rooted graph (G, ρ) is said to admit the *exponential intersection tail* (EIT) property if there exists a measure μ on paths in G, supported only on infinite paths from ρ, with the following property: There exists $0 < \theta < 1$ so that $P_{\mu \times \mu}(|\gamma_1 \cap \gamma_2| > n) < \theta^n$. That is, the probability of two independently picked paths according to μ having more than n edges in common (intersections), decays exponentially in n.

The following Proposition gives an example for a graph admitting the EIT property.

Proposition 4.17. *The EIT property holds for the binary tree.*

Proof. Pick two monotone paths (moving from the root outwards) from the uniform measure on all monotone paths. The probability of the n-th edge being in both paths (i.e. all choices up to the n-th choice are the same) is $(\frac{1}{2})^n$ which decays exponentially in n.   □

Not every graph admits the EIT property, even graphs with $p_c(G) < 1$.

Proposition 4.18. $\mathbb{Z}^2$ *does not have the EIT property.*

Proof. To prove that $\mathbb{Z}^2$ does not satisfy the EIT property, we must show that *any* measure has a significant probability to produce two random paths with many common edges. Let μ by a measure supported on infinite paths starting from the origin. Consider the boundary of the box of size n. What is the probability that two paths picked according to μ have an intersection on the boundary of the box? The square of size n has boundary of size $4n$ and each of the pathes must cross it at least once. Therefore, the probability of the two intersecting on the box boundary is greater than $\frac{1}{4n}$. Indeed, for a point on the boundary of the box of size n denote by $\tilde{\mu}_n(x)$ the probability the first hit of a path distributed according to μ is x, then

$$P_{\mu \times \mu}(\text{The paths intersect in } \partial[-n,n]^2) \geq \sum_{x \in \partial[-n,n]^2} \tilde{\mu}_n(x)^2 \geq \max_x \tilde{\mu}_n(x) \geq \frac{1}{4n}.$$

Thus,

$$\mathbb{E}_{\mu \times \mu}\left[\text{\# of intersections}\right] = \sum_{n=1}^{\infty} \mathbb{E}_{\mu \times \mu}\left[\text{\# of intersections on the square of size } n\right]$$

$$> \sum_{n=1}^{\infty} \frac{c}{n} = \infty.$$

Since this holds for any probability measure μ and the EIT property implies that the last expectation must be finite for the appropriate measure we conclude that for $\mathbb{Z}^2$ the EIT property does not hold.   □

Exercise 4.19.   (Level 3)

1. Show that for $d \geq 4$ the lattice $\mathbb{Z}^d$ admits the EIT property.
2. Show that nonamenable graphs admit the EIT property.

Remark 4.20.   In [BPP98] EIT is established for $\mathbb{Z}^3$.

Exercise 4.21.   (Level 3) Prove that in $\mathbb{Z}^3$ the uniform measure on simple random walk paths does not admit the EIT property. We give several hints: Require the first walk to fill a large $n \times n$ box around the origin, by making it return to the root many times. Make sure that this requirement is satisfied with probability decaying polynomially in n. The other random walk should meet this walk at least n times, inside the $n \times n$ box.

Open problem 4.22. *Does the* loop erased random walk *(LERW) on* $\mathbb{Z}^d$ *admits the EIT property? See [LL10] for the definition of LERW.*

Proposition 4.23. *The uniform measure on monotone random walks, i.e. random walks which can go only in one direction at each coordinate direction, admits the EIT property in* $\mathbb{Z}^d$ *when* $d \geq 4$.

The proof of the above lemma is left as a challenging exercise but can also be found in [BPP98].

Exercise 4.24. (Level 3) Assume that (G, ρ) is a rooted graph with the property: there exists $0 < \theta < 1$ such that $\mathbb{P}\,[\text{SRW on } G \text{ returns to the root after } n \text{ steps}] < \theta^n$. Does the graph admits the EIT property?

The next Theorem is part of Proposition 1.2 from [BPP98].

Theorem 4.25. *If* G *admits the EIT property with parameter* θ, *then* $p_c(G) < 1$. *In fact,* $p_c(G) \leq \theta$.

Proof. Consider edge percolation with parameter p. Let μ be one of the measures satisfying the EIT property with parameter θ. For an infinite path γ, denote by $\gamma[0, n]$ the first n steps (i.e. edges) of the path and define the random variable Z_n by

$$Z_n := \int_\gamma p^{-n} \mathbb{1}_{\{\gamma[0,n] \text{ is open}\}} d\mu.$$

Note that up to the normalization p^{-n} this is exactly the μ measure of the paths that stay in the open cluster of the origin for n steps. Thus it is enough to prove that there exists some $c > 0$ such that $\mathbb{P}_p[Z_n > 0] > c$ for every $n \in \mathbb{N}$. First observe that by Fubini the expectation of Z_n is 1. Using the Paley Zygmond inequality [PZ32] which states that

$$\mathbb{P}_p(Z_n > 0) \geq \frac{\mathbb{E}_p[Z_n]^2}{\mathbb{E}_p[Z_n^2]},$$

and the fact that $\mathbb{E}_p[Z_n] = 1$, it's enough to find a uniform upper bound (in n) on $\mathbb{E}_p[Z_n^2]$. We have

$$
\begin{aligned}
\mathbb{E}_p[Z_n^2] &= \int_{\gamma_1} \int_{\gamma_2} p^{-2n} \mathbb{1}_{\{\gamma_1 \text{ and } \gamma_2 \text{ are open}\}} d\mu(\gamma_1) d\mu(\gamma_2) \\
&\leq \int_{\gamma_1} \int_{\gamma_2} p^{-|\gamma_1 \cap \gamma_2|} d\mu(\gamma_1) d\mu(\gamma_2) \\
&\leq \sum_{i=0}^{\infty} \left(\frac{\theta}{p}\right)^i,
\end{aligned}
\tag{4.1}
$$

where the second inequality follows from the EIT property. Thus for $p > \theta$ the last sum is finite and therefore $\sup_n E[Z_n^2] < \infty$, as required.   □

Exercise 4.26. (Level 2) Using the methods described above show that $p_c(\text{binary tree}) = \frac{1}{2}$, and $p_c(d - \text{regular tree}) = \frac{1}{d}$.

The EIT property concerns rooted paths in a graph. One can consider other rooted subgraphs instead of rooted paths, such as trees or even lattices.

Open problem 4.27. *Is there a measure on embedding of $\mathbb{Z}^2$ into $\mathbb{Z}^d$ for some $d \geq 3$, with an EIT-like property?*

Given two vertices in a graph, constructing a measure on paths between them with a given length and the EIT property, can give a lower bound on the probability they are connected in Bernoulli percolation. The proof of such a bound follows the same lines as the proof of Theorem 4.25

Exercise 4.28. (Level 3) Show that the $[n]^4$ lattice tori admits a connected component of size $\frac{n^4}{2}$ in Bernoulli percolation with positive probability for p sufficiently close to 1, independent of n^4.

Exercise 4.29. (Level 2) Show that the EIT property is invariant under rough isometries between bounded degree graphs. See Chap. 6 for the definition of rough isometry.

Open problem 4.30. *In a three dimensional slab $\{(n, f(n), g(n))\}_{n \in \mathbb{N}}$, under what conditions on f and g there is an EIT measure in the slab?*

4.4 Self Avoiding Walk

Connective constants for self avoiding walks admit partial analogies with the critical probability of percolation. Both are monotone with respect to inclusion and graph covering. With the help of Hugo Duminil-Copin we briefly mention a couple of conjectures. For more details on the model see [BDCGS12, MS93] and the very recent [GL13] regarding SAW's on vertex transitive graphs.

Let G be a graph, usually we discuss a transitive graph, but for concreteness one can think on $\mathbb{Z}^d$. Self avoiding walk (SAW) is a measure on random walks which is supported on paths that do not return to a vertex that already been visited. More precisely we define,

Definition 4.31.

- *SAW(n)* is the uniform measure on all self avoiding paths of length n starting from a fixed vertex.
- *SAL(n)* is the uniform measure on self avoiding loops (i.e. closed self avoiding paths) of length n starting from a fixed vertex.

  $\mu = \lim |SAW(n)|^{1/n}$ is the connective constant. Let $\mu_{loops} = \lim |SAL(n)|^{1/n}$.

Conjecture 4.32. Let G be a vertex transitive graph with $h(G) > 0$. Then $\mu_{loops} < \mu$.

Except for very small family of amenable vertex transitive graphs such as the ladder, we expect $\mu_{loops} = \mu$. This is the self avoiding walk analogue of Kesten's amenability criterion.

We don't know how to show that SAW on planar hyperbolic lattices is ballistic.

Conjecture 4.33. μ is continuous with respect to local convergence of infinite vertex transitive graphs.

See the next Chap. 5 for local convergence. Another conjecture on SAW is the following:

Conjecture 4.34. $I - dim(G) > 1$ implies $\mu > 1$.

Given a Cayley graph, for any generating set corresponds a connective constant μ. This suggests a canonical generating sets minimizing μ. Gady Kozma conjectured that for planar Cayley graphs μ is algebraic, and he showed that the set of all connective constants of groups contains an interval.

Chapter 5
Local Limits of Graphs

5.1 The Local Metric

In this section we will only consider connected simple graphs (i.e. without loops or multiple edges). We start by recalling few definitions from previous sections. If $G = (V, E)$ is such a graph, and $x, y \in V$, the graph distance between x and y in G is defined to be the length of a shortest path in G between x and y, and is denoted by $d_G(x, y)$. A *rooted* graph (G, ρ) is a graph G together with a distinguished vertex ρ of G. Two rooted graphs (G, ρ) and (G', ρ') are said to be equivalent, a notion denoted by $(G, \rho) \simeq (G', \rho')$ if there is a graph homomorphism $G \to G'$ that maps ρ to ρ'. Following [BS01b], we define a pseudo-metric on the set of all locally finite connected rooted graphs by

$$
d_{loc}\left((G, \rho), (G', \rho')\right) = \left(1 + \sup\left\{n \geq 0 : (B_G(\rho, n), \rho) \simeq (B_{G'}(\rho', n), \rho')\right\}\right)^{-1},
$$
(5.1)

where $B_G(\rho, n)$ stands for the combinatorial ball of radius n around ρ in G. Hence d_{loc} induces a metric (still denoted d_{loc}) on the set $\mathcal{G}_\bullet$ of equivalence classes of (locally finite connected) rooted graphs.

Proposition 5.1. *The metric space $(\mathcal{G}_\bullet, d_{loc})$ is Polish (separable and complete). Furthermore for every $M > 0$ the subspace $\mathcal{G}_\bullet^M \subset \mathcal{G}_\bullet$ of (isometry classes of) graphs with maximal degree bounded by M is compact.*

Remark 5.2. Formally, elements of $\mathcal{G}_\bullet$ are equivalence classes of rooted graphs, but we will not distinguish between graphs and their equivalence classes and therefore, with some abuse of notation, will use the same terminology and notation in both cases. One way to bypass this identification is to choose once and for all a canonical representant in each class, see [AL07, Sect. 2] for details.

Proof. The countable set of finite rooted graphs is dense in $(\mathcal{G}, d_{loc})$. If (G_n, ρ_n) is a Cauchy sequence for d_{loc} it is easy to see that the combinatorial balls $B_{G_n}(\rho_n, r)$

I. Benjamini, *Coarse 3Geometry and Randomness*, Lecture Notes in Mathematics 2100, DOI 10.1007/978-3-319-02576-6_5,
© Springer International Publishing Switzerland 2013

rooted at ρ_n are stationary in n for all r and thus converge to some $B_G(\rho, r)$ for some (possibly infinite) graph (G, ρ) which is checked to be the limit of the G_n's. For the second assertion, notice that there are only finitely many rooted graphs in $\mathcal{G}_\bullet^M$ of radius r. Hence a diagonal procedure yields the result.   □

Obviously, if G is a vertex transitive graph then all the balls of radius r around any vertex of G are isomorphic and thus it makes sense to speak about convergence of (non-rooted) transitive graph in the sense of d_{loc}, more precisely we say that a sequence (G_n) of vertex transitive graphs is locally converging towards (a vertex transitive) graph G, if for any $\rho_n \in G_n$ and $\rho \in G$ the rooted graphs (G_n, ρ_n) are converging towards (G, ρ) with respect to d_{loc}.

Example 5.3. The $n \times n$-grid tori converge towards $\mathbb{Z}^2$ as $n \to \infty$.

Example 5.4. Let G_n be a sequence of d-regular graphs such that girth$(G_n) \to \infty$ as $n \to \infty$. Then G_n converges towards a d-regular tree with respect to d_{loc}.

Definition 5.5. We say that an infinite vertex transitive graph G is $f(r)$-sofic, if for any r, the minimum of the diameters of finite vertex transitive graph with the same r ball as in G, is $f(r)$.

Open problem 5.6. *Can you find a counter example to the speculation that if a graph is $f(r)$-sofic for some $f(r)$, then there is $c < \infty$ such that it is cr sofic? Moreover, is there a constant $c(d) < \infty$ such that, if $B(0, r)$ appears as a r-ball in a d-regular finite vertex transitive graph, then there is a vertex transitive graph with diameter at most $c(d)r$ admitting $B(0, r)$ as an r-ball? How small $c(d)$ can be? Note that for trees this is the girth problem.*

5.2 Locality of p_c

Theorem 5.7 (Generalization of Theorem 11.10). *Let $\{G_n\}_{n \geq 1}$ be an expander family. Let G be a graph such that $G_n \to G$ with respect to d_{loc} as $n \to \infty$. Then:*

1. If $p > p_c(G)$ then there exists $\alpha > 0$ such that

$$\mathbb{P}_p \left(\begin{array}{c} \text{there is an open cluster} \\ \text{in } G_n \text{ of size} \geq \alpha|G_n| \end{array} \right) \xrightarrow[n \to \infty]{} 1.$$

2. If $p < p_c(G)$ then for every $\alpha > 0$

$$P_p \left(\begin{array}{c} \text{there is an open cluster} \\ \text{in } G_n \text{ of size} \geq \alpha|G_n| \end{array} \right) \xrightarrow[n \to \infty]{} 0.$$

Proof. The proof is very similar to the proof of related theorem on expanders and can be found in [BNP11]. □

The question we wish to discuss here is whether $p_c(G)$ is a local property? More precisely, if $G_n \to G$ with respect to d_{loc} as $n \to \infty$, does this imply also that $p_c(G_n) \to p_c(G)$?

In general this is not always the case as shown by the following example:

Example 5.8. The graphs $\mathbb{Z} \times \mathbb{Z}/n\mathbb{Z}$ converge locally towards $\mathbb{Z}^2$, but for every $n \geq 0$ $p_c(\mathbb{Z} \times \mathbb{Z}/n\mathbb{Z}) = 1$ which is different from $p_c(\mathbb{Z}^2)$.

However, under some additional assumptions we believe the answer to the last question is positive:

Conjecture 5.9. Assume there exists $c < 1$ such that for every $n \geq 0$ we have $p_c(G_n) < c$. Show that if $G_n \to G$, then $p_c(G_n) \to p_c(G)$ as $n \to \infty$.

Another "easier" open problem is:

Open problem 5.10. *Show that if $\{G_n\}_{n \geq 1}$ is an expander family and $G_n \to G$ with respect to d_{loc}, then $p_c(G_n) \to p_c(G)$. Note that if the previous conjecture is true, this open-problem will follow from it (since the condition G_n is an expander family implies that $p_c(G_n) < c$ for some $c < 1$).*

In [BNP11] it was proved that:

Theorem 5.11. *Let $\{G_n\}_{n \geq 1}$ be a d-regular expander family with the property that girth$(G_n) \to \infty$ as $n \to \infty$. Then $p_c(G_n)$ converges to $\frac{1}{d-1} = p_c(d$-regular tree).*

Another particular case of the above conjecture that was proved by Grimmett and Marstrand (see [Gri99]) is the following:

Theorem 5.12 (Grimmett and Marstrand). *For $d > 1$*

$$p_c\left(\mathbb{Z}^d \times (\mathbb{Z}/n\mathbb{Z})^k\right) \xrightarrow[n \to \infty]{} p_c(\mathbb{Z}^{d+k}).$$

A related but not directly linked fact is the so-called mean-field behavior of the percolation in very high dimensions which in particular implies that the critical probability come close to that of a tree with the same degree.

Theorem 5.13. *We have [HS90],*

$$p_c(\mathbb{Z}^d) = \frac{1}{2d} + o\left(\frac{1}{d}\right) \text{ as } d \to \infty.$$

Intuitive explanation of the last theorem is the following: every vertex locally "feels" like in a $2d$-regular tree, as the number of crossing paths of a given length is negligible compared to the total number of paths of the same length, for a rather simple proof without the sharp error terms see [ABS04].

5.3   Unimodular Random Graphs

The metric d_{loc} enables us to speak about convergence of rooted graphs. The root vertices of these graphs play a crucial role in that convergence. In the following, the idea is to randomize the root vertex in order to get a typical view on a given graph. This has lead to the concept of unimodular random graphs. For more details, we refer to [AL07].

The Mass-Transport Principle (continuous version)

Let (G, ρ) be a random (rooted) graph which is almost surely finite according to some law μ. Conditioned on (G, ρ) choose independently a new root $\tilde{\rho}$ uniformly from the vertices of G. Then, the distribution v of $(G, \tilde{\rho})$ satisfies

$$\int f(G, \tilde{\rho}) dv(G, \tilde{\rho}) = \int \frac{1}{|G|} \sum_{x \in G} f(G, x) d\mu(G, \rho),$$

for any positive Borel function f.

If the random graph $(G, \tilde{\rho})$ has the same distribution as (G, ρ), that is $\mu = v$, we say that (G, ρ) is *unbiased* or *uniformly rooted*.

Example 5.14. Let (G_n, ρ_n) be the full binary tree up to level n with a uniformly root vertex ρ_n. Then, the local limit of (G_n, ρ_n) **is not** the full binary tree as seen from a root, but rather the full binary tree as seen from the top. More precisely this is the canopy tree (see Example 1.58) rooted at a point of level n with probability $2^{-(n+1)}$.

The class of unbiased random graphs is particulary interesting and generalizes Cayley graphs since they satisfy a generalized version of the *The Mass-Transport-Principle* (see Lemma 8.14). Suppose we are given a function f that takes as parameters a graph G and two vertices $x, y \in G$ and returns a non-negative number. Suppose also that f is homomorphism invariant that is, if $T : G \rightarrow G'$ is a graph homomorphism that takes x to x' and y to y' then $f(G, x, y) = f(G', x', y')$. Finally assume that f is measurable in the space $\mathcal{G}_{\bullet\bullet}$ of isometry classes of bi-rooted graphs (with an easy extension of d_{loc}).

Definition 5.15 ([AL07, BS01b]). A random rooted graph (G, ρ) with distribution μ is said to satisfy the (generalized) Mass-Transport-Principle (MTP) if

$$\int_{\mathcal{G}_\bullet} \sum_{x \in G} f(G, \rho, x) d\mu(G, \rho) = \int_{\mathcal{G}_\bullet} \sum_{x \in G} f(G, x, \rho) d\mu(G, \rho), \qquad (5.2)$$

such a random graph is also called *unimodular* (coming from the terminology of transitive graphs).

Later such functions f will be called *mass-transport functions* and will be interpreted as a quantity of mass that x sends to y. Thus the MTP principle implies that the mean quantity of mass that the root ρ sends is equal to the mean quantity it receives.

Remark 5.16. Note that the sum over $x \in G$ in (5.2) has no meaning because (G, ρ) is formally an equivalence class of rooted graphs and not a specific rooted graph. However it is easily checked that the quantity we are interested in does not depend on a representative of (G, ρ).

Claim 5.17. *Cayley graphs and uniformly rooted random graphs satisfy MTP.*

Proof. Let (G, ρ) be distributed according to μ a uniformly rooted random graph. We have

$$\int_{\mathcal{G}_\bullet} \sum_{y \in G} f(G, \rho, y) d\mu(G, \rho) = \int_{\mathcal{G}_\bullet} \frac{1}{|G|} \sum_{x, y \in G} f(G, x, y) d\mu(G, \rho)$$

$$= \int_{\mathcal{G}_\bullet} \frac{1}{|G|} \sum_{x, y \in G} f(G, y, x) d\mu(G, \rho)$$

$$= \int_{\mathcal{G}_\bullet} \sum_{y \in G} f(G, y, \rho) d\mu(G, \rho).$$

$\square$

Exercise 5.18. (Level 2) Show that a random rooted finite graph which satisfies the MTP is uniformly rooted.

The definition of uniformly rooted graph cannot be made precise for infinite random graphs. However the MTP still has sense in that setting. And we have:

Proposition 5.19 ([BS01b]). *Let (G_n, ρ_n) be a sequence of unimodular random graphs that converge in distribution with respect to d_{loc} towards (a possibly infinite) random rooted graph (G, ρ). Then (G, ρ) satisfies the MTP.*

Proof. Suppose first that f is a mass-transport-function that only depends on a finite neighborhood around the root vertex that is $f(G, x, y) = \mathbf{1}_{\{d_G(x, y) \le k\}} f(B_G(x, k), x, y)$ for some $k \ge 0$. For $l \in \{1, 2, \ldots\} \cup \{\infty\}$ denote

$$f_l(G, x, y) = \mathbf{1}_{\{\#B_G(x, k) < l\}} \mathbf{1}_{\{\#B_G(y, k) < l\}} \big(f(G, x, y) \wedge l \big),$$

where (G, x, y) is any bi-rooted graph. Define the functionals

$$\psi_l : (G, \rho) \in \mathcal{G}_\bullet \longmapsto \sum_{x \in G} f_l(G, \rho, x)$$

$$\phi_l : (G, \rho) \in \mathcal{G}_\bullet \longmapsto \sum_{x \in G} f_k(G, x, \rho),$$

and notice that they are bounded and continuous thanks to the fact that f only depends on a finite neighborhood. Since (G_n, ρ_n) are unimodular we have $\mathbb{E}[\psi_l(G_n, \rho_n)] = \mathbb{E}[\phi_l(G_n, \rho_n)]$ and thus $\mathbb{E}[\psi_l(G, \rho)] = \mathbb{E}[\phi_l(G, \rho)]$. By monotone convergence we get $\mathbb{E}[\psi_\infty(G, \rho)] = \mathbb{E}[\phi_\infty(G, \rho)]$ which is the MTP for the function f. Going from functions satisfying the above hypothesis to general ones is a standard (but a bit technical) argument.   □

Open problem 5.20 ([AL07]). *Let (G, ρ) be an (infinite otherwise it is easy) unimodular random graph. Does there exist a sequence (G_n, ρ_n) of finite uniformly rooted graphs such that $(G_n, \rho_n) \to (G, \rho)$ in distribution with respect to* d_{loc}*?*

Stationary Along SRW

We now give another point of view about unimodular random graphs as, roughly speaking, random rooted graphs whose distribution is invariant under re-rooting along a simple random walk path.

Consider a mass-transport function which satisfy $f(G, x, y) = f(G, x, y)\mathbf{1}_{x \sim y}$, where $x \sim y$ means that x and y are neighbors in the graph G. Applying the MTP to a unimodular random graph (G, ρ) with law μ and function f as above we get

$$\int_{\mathcal{G}_\bullet} \sum_{x \sim \rho} f(G, \rho, x) d\mu(G, \rho) = \int_{\mathcal{G}_\bullet} \sum_{x \sim \rho} f(G, x, \rho) d\mu(G, \rho),$$

or equivalently

$$\int_{\mathcal{G}_\bullet} \deg(\rho) \frac{1}{\deg(\rho)} \sum_{x \sim \rho} f(G, \rho, x) d\mu(G, \rho) = \int_{\mathcal{G}_\bullet} \deg(\rho) \frac{1}{\deg(\rho)}$$

$$\sum_{x \sim \rho} f(G, x, \rho) d\mu(G, \rho).$$

In other words, if $(\bar{G}, \bar{\rho})$ is distributed according to (G, ρ) biased by $\deg(\rho)$ (assuming that $\int d\mu \; \deg(\rho) < \infty$) and if conditioned on $(\bar{G}, \bar{\rho})$, X_1 is distributed as the first step of a simple random walk starting from $\bar{\rho}$ in $\bar{G}$ then we have the following equality in distribution

$$(\bar{G}, \bar{\rho}, X_1) \stackrel{(d)}{=} (\bar{G}, X_1, \bar{\rho}). \tag{5.3}$$

A graph $(\bar{G}, \bar{\rho})$ satisfying the last property is called *stationary and reversible*. It is also possible to go from a stationary and reversible random graph towards a unimodular random one by biasing by $\deg(\rho)^{-1}$, see [AL07, BC12].

Another way to introduce the same concept is to say that if we consider a random graph (G, ρ) and let a bi-infinite simple random walk $\{X_n\}_{n \in \mathbb{Z}}$ run over it, such that $X_0 = \rho$. Then we have a probability distribution over the set of (equivalence classes of) graphs with a bi-infinite path on them. The stationarity and reversibility assumption tells us that this probability distribution is invariant under the shift operations which consist of translating the root point of the path by 1 or -1. Hence much of ergodic theory can be applied.

Measured Equivalence Relations

The MTP has also many connection with so-called measured equivalence relations. We will not enter the details and refer to [AL07] or [BC12] for more details. Let us just show that there is a dictionary between the two notions:

Measured graphed equivalence relation	Random rooted graph
Harmonic	Stationary
Totally invariant	Reversible
Measure preserving	Unimodular

5.4 Applications

We give here as illustrations and motivations some known results about unimodular random graphs.

Theorem 5.21 ([BLPS99a]). *Let (G, ρ) be an a.s. infinite unimodular random graph. Then the expected degree of the root is bigger than 2.*

Proof. Let us consider the mass-transport function $f(G, x, y)$ which sends a unit of mass from x to y if there is exactly one edge between x and y (in particular they must be neighbors) and the removal of this edge leaves x in a finite component. Then applying the MTP to show that the expected mass that ρ sends is equal to the expected mass it receives. Let us examine the different cases.

- If $\deg(\rho) = 1$ then ρ sends mass 1 to its only neighbor and receives nothing.
- If $\deg(\rho) \geq 2$ and if ρ sends his mass to somebody then the mass it receives is less than $\deg(\rho) - 1$.
- If $\deg(\rho) \geq 2$ and if ρ does not send his mass then the mass it receives is less than $\deg(\rho) - 2$.

In any case $D + S - R \geq 2$ with obvious notation. Taking expectation yields the result.   □

The following theorem is very closely related to Theorem 1.49

Theorem 5.22 ([LS99]). *Let (G, ρ) be a unimodular random graph then the number of ends of G is either $0, 1, 2$ or ∞.*

Proof Sketch. Let D_n be the set of vertices v of (G, ρ) such that the removal of $B_G(v, n)$ disconnects G into at least three infinite components. We consider the mass-transport function $f(G, x, y)$ which sends a unit of mass from x and distributes it equally among all vertices z such that x is in an infinite connected component of $G \backslash B_G(y, n)$ where y is a closest point of D_n to x with respect to the graph metric d_G and $d_G(y, z) \leq l$, ($f = 0$ otherwise). Applying the MTP we get

$$\int_{\mathcal{G}_\bullet} \sum_{x \in G} f(G, \rho, x) \mathrm{d}\mu(G, \rho) = \int_{\mathcal{G}_\bullet} \sum_{x \in G} f(G, x, \rho) \mathrm{d}\mu(G, \rho). \qquad (5.4)$$

The left-hand side is the quantity of mass that ρ sends, it is thus less than 1. Hence the right-hand side, which is the mean quantity of mass that ρ receives, is finite as well. Imagine now that G has an isolated end and more than three ends, with positive probability. Then there exists a finite set that disconnects this end from the other ones and for some n and l the root itself will get an infinite mass.   □

The Kaimanovich-Vershik entropic criterion (see Sect. 12) can also be generalized to unimodular random graphs. This leads to:

Theorem 5.23 ([BC12]). *Let (G, ρ) be a unimodular random graph satisfying*

- $\mathbb{E}[\deg(\rho)] < \infty$,
- $\mathbb{E}[\log(B_G(\rho, r))] = o(r)$.

Then (G, ρ) is almost surely Liouville (that is admits no non-constant bounded harmonic functions).

Theorem 5.24 ([BS01b]). *Let (G_n, ρ_n) be a sequence of uniformly rooted finite random graphs with the following properties:*

(i) there exists $M > 0$ such that $\sup_n \sup_{v \in G_n} \deg(v) < M$,
(ii) the random graphs (G_n, ρ_n) are a.s. simple planar graphs.

If $(G_n, \rho_n) \to (G, \rho)$ in distribution with respect to $\mathrm{d}_{\mathrm{loc}}$ then (G, ρ) is almost surely recurrent.

5.5   Some Examples

In this subsection we present several examples of unimodular random graphs, some of them come from slight random modifications of transitive graphs and others are purely random.

Construction From Existing (Random) Graphs

Let (G, ρ) be a unimodular random graph. Condition on (G, ρ), consider a Bernoulli percolation of parameter r and denote $\mathcal{C}(\rho)$, the connected component of ρ after applying the percolation process. We already used a variant of the following:

Proposition 5.25. *The random rooted graph* $(\mathcal{C}(\rho), \rho)$ *is unimodular.*

Proof. We directly verify the MTP. Denote μ the distribution of (G, ρ) and let f be a transport function. To simplify notation, we write $\mathcal{C}$ instead of $\mathcal{C}(\rho)$. We have

$$\int \sum_{x \in \mathcal{C}} f(\mathcal{C}, \rho, x) d\mathbb{P}(\mathcal{C}, \rho) = \int \sum_{x \in G} \underbrace{\int f(\mathcal{C}, \rho, x) \mathbf{1}_{x \in \mathcal{C}} d\mathbb{P}(B_e)_{e \in E(G)}}_{F(G, \rho, x)} d\mu(G, \rho).$$

Thus the function $F(., ., .)$ is a transport function and applying the MTP for the original unimodular graph yields the result. □

Remark 5.26. We just used a weak property of the Bernoulli percolation in order to say that F is a transport function. This reasoning is valid for any invariant percolation process.

Exercise 5.27. (Level 1) Let (G, ρ) be a unimodular random graph. Delete all vertices of degree at least M for some $M \geq 0$. Show that the component of the root vertex rooted at ρ is still a unimodular random graph.

Augmented Galton-Watson Trees (AGW)

Let $\mathbf{p} = (p_k)_{k \geq 0}$ be a probability distribution over $\mathbb{Z}_+$. A Galton-Watson tree is a random rooted tree (which in fact bears an additional planar structure that we forget in our setting) defined informally as follows: Start with the root ρ of the tree and sample its number of children according to $\mathbf{p}$, then iterate independently for each child obtained. The random rooted tree obtained is not homogeneous because the root has stochastically one neighbor less than the other vertices. To cope up with this phenomenon, we define the *Augmented* Galton-Watson measure as the measure obtained when we force the root ρ to have one more child that is when its offspring distribution over $\mathbb{Z}_+$ is given by $\mathbb{P}(\rho \text{ has } k \text{ children}) = p_{k-1}$ for $k \geq 1$.

The random rooted tree (T, ρ) obtained by the above device is not unimodular, indeed consider the transport function $f(T, x, y) = \frac{1}{\deg(x)}$ if $x \sim y$ and 0 otherwise. If it held, the MTP would give

$$1 = \int_{\mathcal{G}_\bullet} d\mu(T, \rho) \sum_{x \sim \rho} \frac{1}{\deg(x)} = \mathbb{E}[1 + X]\mathbb{E}\left[\frac{1}{1 + X}\right], \tag{5.5}$$

where X is a random variable over $\mathbb{Z}_+$ distributed according to **p**. However, the last inequality is not always fulfilled.

Theorem 5.28 ([LPP95]). *The random variable (T, ρ) biased by $\deg^{-1}$ is unimodular.*

Proof sketch. We will not give all the details of this proof but only indicate why we have to bias by $\deg(\rho)^{-1}$. In fact the random variable (T, ρ) is a stationary and reversible random graph. This can be seen heuristically as follows. Imagine that we start at the root ρ of T and that we take a one step random walk. The scenery that we see is an edge that we just came from, and two independent GW trees grafted on that edge, which is the same as (T, ρ) in distribution. By the discussing of the previous section , (T, ρ) biased by $\deg(\rho)^{-1}$ is unimodular.   □

To appreciate the relevance of unimodularity over transitivity see [Häg11].

Example 5.29. The product of two critical Galton-Watson trees conditioned to survive gives a unimodular random graph.

Exercise 5.30. (Level 4)

- Show that the last example is transient yet subdiffusive.

- More precisely, show that the return probability at time n is roughly $\left(n^{-\frac{2}{3}}\right)^2$, i.e., the diffusion constant is $\frac{1}{3}$.

5.6 Growth and Subdiffusivity Exponents

In the following we will construct a unimodular random rooted tree (T, ρ) such that the volume growth around the root is of order r^α for $\alpha \geq 1$ thus proving that any growth exponent can arise in the theory of unimodular random graph.

Fix $\alpha > 1$. We consider the sequence $\epsilon_1, \ldots, \epsilon_n, \ldots \in \{1, 2\}$ defined recursively as follows: Start with $\epsilon_1 = 1$, if $\epsilon_1, \ldots, \epsilon_k$ are constructed we let $\xi_k = \prod_{i=1}^k \epsilon_k$, and set $\epsilon_{k+1} = 1$ if $\xi_k > k^\alpha$ and $\epsilon_{k+1} = 2$ otherwise. Clearly there exist constants $0 < c < C < \infty$ such that $ck^\alpha \leq \xi_k \leq Ck^\alpha$ for every $k \geq 1$. We now consider the tree T_n of height n, starting from an initial ancestor at height 0 such that each vertex at height $0 \leq k \leq n - 1$ has ϵ_{n-k} children. Hence the tree T_n has only simple or binary branchings. The *depth* $D(u)$ of a vertex u in T_n is defined to be n minus its height. For example the leaves of T_n have depth 0. We also define the depth of an edge as the maximal depth of its ends. If u is a leaf of T_n then for every $0 \leq r \leq n$, the ball of radius r around u in T_n is contained in the set of descendants of the ancestor of u at depth r. This subtree has precisely $\sum_{i=0}^r \frac{\xi_r}{\xi_{r-i}}$ vertices (with the convention $\xi_0 = 1$) so we deduce that

$$\#B_{T_n}(u,r) \le \sum_{i=0}^{r} \frac{\xi_r}{\xi_{r-i}} \le C'r^\alpha, \qquad (5.6)$$

for some C' independent of r. It is easy to see that the last bound still holds for any vertex $u \in T_n$ (not necessarily a leaf) provided that C' is made large enough. We also introduce the tree T_∞ which is composed of an infinite number of vertices at depth $0, 1, 2, 3, \ldots$ such that each vertex at depth k is linked to ϵ_{k+1} vertices at depth $k - 1$. We transform these graphs into random rooted ones by choosing the root ρ_n uniformly among all vertices of T_n.

Proposition 5.31. *We have the convergence in distribution with respect to* d_{loc}

$$(T_n, \rho_n) \xrightarrow[n\to\infty]{} (T_\infty, \rho), \qquad (5.7)$$

for a particular choice of a random root $\rho \in T_\infty$. *In particular* (T_∞, ρ) *is an unimodular random tree with one end such that there exist* $0 < c_1 < c_2 < \infty$ *with*

$$c_1 r^\alpha \le \mathbb{E}\left[\#B_T(\rho, r)\right] \le c_2 r^\alpha.$$

Proof. It is enough to show that $D(\rho_n)$ converges in distribution to a non degenerate random variable denoted by D as $n \to \infty$. Indeed if we choose a random root $\rho \in T_\infty$ with depth given by D, since the r-neighborhood of a vertex at depth k in T_n and in T_∞ are the same when $n \ge r + k$, we easily deduce the weak convergence of (T_n, ρ_n) to (T_∞, ρ) for d_{loc}. Furthermore since (T_n, ρ_n) are unbiased random graph (T, ρ) will automatically be unimodular.

Let $k \ge 0$. The probability that $D(\rho_n) = k$ is exactly the proportion of vertices in T_n at depth k which is

$$\frac{\xi_n/\xi_k}{\sum_{i=0}^{n} \xi_n/\xi_i} \xrightarrow[n\to\infty]{} \frac{\xi_k^{-1}}{\sum_i \xi_i^{-1}}.$$

The series $\sum \xi_i^{-1}$ is convergent since $\xi_i \sim i^{-\alpha}$. We easily deduce that $D(\rho_n)$ converges in distribution when $n \to \infty$. The last part of the theorem follows from the remarks made on the volume growth inside T_∞. $\qquad \square$

Exercise 5.32. (Level 2) Let $(X_n)_{n\ge0}$ be a simple random walk on (T, ρ) and for $r \ge 0$ denote τ_r the first time the walk reach distance r from the root. Show that $\kappa_1 r^{1/\alpha} \le \mathbb{E}[\tau_r] \le \kappa_2 r^{1/\alpha}$ for some $0 < \kappa_1 < \kappa_2 < \infty$.

Chapter 6
Random Planar Geometry

What is a typical random surface? This question has arisen in the theory of two-dimensional quantum gravity where discrete triangulations have been considered as a discretization of a random continuum Riemann surface. As we will see the typical random surface has a geometry which is very different from the one of the Euclidean plane.

6.1   Uniform Infinite Planar Triangulation (UIPT)

A *planar map* is an embedding of a finite connected planar graph into the two-dimensional sphere up to continuous deformations that preserve the orientation. We deal with planar maps because the little additional structure they bear compared to planar graphs enable us to do combinatorics with them more easily. A planar map is called a triangulation if all its faces have degree three and is called rooted if it has a distinguished oriented edge. We denote $\mathcal{T}_n$ the set of all rooted triangulations with n faces.

The following theorem defines the model of Uniform Infinite Planar Triangulation (UIPT):

Theorem 6.1 ([AS03]). *Let T_n be uniformly distributed over $\mathcal{T}_n$ and let (T_n, ρ) (with a slight abuse of notation) be its associated graph rooted at the origin of the root edge of T_n then we have the following convergence in distribution with respect to $\mathrm{d}_{\mathrm{loc}}$*

$$(T_n, \rho) \xrightarrow[n \to \infty]{} (T_\infty, \rho), \tag{6.1}$$

I. Benjamini, *Coarse 3Geometry and Randomness*, Lecture Notes in Mathematics 2100,
DOI 10.1007/978-3-319-02576-6_6,
© Springer International Publishing Switzerland 2013

where (T_∞, ρ) is a random infinite rooted planar graph called the Uniform Infinite Planar Triangulation (UIPT).[1]

The geometry of UIPT is very interesting and far from the Euclidean one. For examples, Angel showed [Ang03] that the typical volume of a ball of radius r in UIPT is of order r^4. This random graph (and its family) has been extensively studied over the last ten years, see the works of Angel and Schramm, Chassaing and Durhuus, Krikun, Le Gall and Ménard... See also impressive work of Le Gall and Miermont on a different but related point of view: Scaling limits of random maps.

Remark 6.2. UIPT is in fact a stationary and reversible random graph, hence its biased version by $\deg(\rho)^{-1}$ is unimodular. See [AS03].

Unfortunately (or perhaps fortunately?) basic questions about UIPT are still open. Here is the most basic one:

Conjecture 6.3 ([AS03]). The simple random walk on UIPT is recurrent.

Added in proofs: just solved by Gurel-Gurevich and Nachmias [GG13].

In [BC13] it is shown that the simple random walk on the related Uniform Infinite Planar Quadrangulation (UIPQ) is subdiffusive with exponent less that $1/3$.

Conjecture 6.4 ([BC13]). The simple random walk $\{X_n\}_{n \geq 0}$ on the UIPT is subdiffusive with exponent $1/4$, i.e.

$$d_{gr}(X_0, X_n) \asymp n^{1/4}.$$

6.2   Circle Packing

Since random triangulations and UIPT are planer graphs, it is very tempting to try and understand their conformal structures. The theory of Circle Packing is well-suited for this purpose.

A circle packing on the sphere is an arrangement of circles on a given surface (in our case the sphere) such that no overlapping occurs and so that all circles touch another. The most standard question regarding circle packing is there density, i.e., the portion of surface covered by them. The contact graph of a circle packing is defined to be the graph with set of vertices which correspond to the set of circles and an edge between two circles if and only if they are tangent.

Let T_n^S be the set of all triangulations of the sphere $\mathbb{S}_2$ with n faces with no loops or multiple edges. We recall the well known circle packing theorem (see Wikipedia, [HS95]):

---

[1]The real theorem actually deals directly with maps.

Theorem 6.5 (Circle Packing Theorem). *If T is a finite triangulation without loops or multiple edges then there exists a circle packing $P = \{P_c\}_{c \in C}$ in the sphere $\mathbb{S}_2$ such that the contact graph of P is T. In addition this packing is unique up to Möbius transformations.*

Recall that the group of Möbius transformations $z \mapsto \frac{az+b}{cz+d}$, where $a, b, c, d \in \mathbb{C}$ and $ad - bc \neq 0$ can be identified with $\mathrm{PSL}_2(\mathbb{C})$ and that it acts transitively on triplets (x, y, z) of $\mathbb{S}_2$. The circle packing enables us to take a "nice" representation of a triangulation $T \in \mathcal{T}_n$, nevertheless the non-uniqueness is somehow disturbing because to fix a representation we can, for example, fix the images of three vertices of a distinguished face of T. This specification breaks all the symmetry, because sizes of some circles are chosen arbitrarily. Here is how to proceed:

The action on $\mathbb{S}_2$ of an element $\gamma \in \mathrm{PSL}_2(\mathbb{C})$ can be continuously extended to $\mathbb{B}_3 := \{(x, y, z) \in \mathbb{R}^3, x^2 + y^2 + z^2 \leq 1\}$: this is the Poincaré-Beardon extension. We will keep the notation γ for transformations $\mathbb{B}_3 \to \mathbb{B}_3$. The action of $\mathrm{PSL}_2(\mathbb{C})$ on $\mathbb{B}_3$ is now transitive on points. The group of transformations that leave 0 fixed is precisely the group $\mathrm{SO}_2(\mathbb{R})$ of rotations of $\mathbb{R}^3$.

Theorem 6.6 (Douady-Earle). *Let μ be a measure on $\mathbb{S}_2$ such that #supp$(\mu) \geq 2$. Then we can associate to μ a "barycenter" denoted by $\mathrm{Bar}(\mu) \in \mathbb{B}_3$ such that for all $\gamma \in \mathrm{PSL}_2(\mathbb{C})$ we have*

$$\mathrm{Bar}(\gamma^{-1}\mu) = \gamma(\mathrm{Bar}(\mu)).$$

We can now describe the renormalization of a circle packing. If P is a circle packing associated to a triangulation $T \in \mathcal{T}_n^S$, we can consider the atomic measure μ_P formed by the Dirac's at tangency point of the disks in P

$$\mu_P := \frac{1}{\#\text{tangency points}} \sum_{\substack{x \text{ is a tangency} \\ \text{point}}} \delta_x.$$

By transitivity there exists a conformal map $\gamma \in \mathrm{PSL}_2(\mathbb{C})$ such that $\mathrm{Bar}(\gamma^{-1}\mu_P)=0$. The renormalized circle packing is by Definition $\gamma(P)$, this circle packing is unique up to rotation of $\mathrm{SO}_2(\mathbb{R})$, we will denote it by $\mathbf{P}_T$. This constitutes a canonical discrete conformal structure for the triangulation.

Here are some open problems regarding circle packing on the sphere:

Open problem 6.7. *If T_n is a random variable distributed uniformly over the set $\mathcal{T}_n^S$, then the variable $\mu_{\mathbf{P}_{T_n}}$ is a random probability measure over $\mathbb{S}_2$ seen up to rotations of $\mathrm{SO}_2(\mathbb{R})$. By classical arguments there exist weak limits μ_∞ of $\mu_{\mathbf{P}_{T_n}}$.*

1. *(Schramm) Determine coarse properties (invariant under $\mathrm{SO}_2(\mathbb{R})$) of μ_∞, e.g. what is the dimension of the support? Start by showing singularity.*

2. *Uniqueness (in law) of μ_∞? In particular can we describe μ_∞ in terms of the Gaussian Free Field? Is it $\exp((8/3)^{1/2}GFF)$, does KPZ hold? See [dup] for more details.*
3. *The random measure μ_∞ can come together with d_∞ a random distance on $\mathbb{S}_2$. Can you describe links between μ_∞ and d_∞? Does one characterize the other?*

6.3   Stochastic Hyperbolic Infinite Quadrangulation (SHIQ)

Recently Guth et al. [GPY11] studied pants decomposition of random surfaces chosen uniformly in the moduli space of hyperbolic metrics equipped with the Weil-Peterson volume form and a combinatorial analogue obtained by randomly gluing Euclidean triangles (with unit side length) together. They showed that such a random compact surfaces with no genus restriction have large pants decomposition, growing with the volume of the surface. This suggests that the injectivity radius around a typical point is growing to infinity. Gamburd and Makover [GM02] showed that as N grows the genus will converge to $N/4$ and using the Euler's characteristic the average degree will grow to infinity.

Take a uniform measure on triangulations with N triangles conditioned on the genus to be CN for some fixed $C < 1/4$, then we *conjecture* that as N grows to infinity the random surface will locally converge in the sense of [BS01b] (see Sect. 5 above) to a random triangulation of the hyperbolic plane with average degree $\frac{6}{1-4C}$. In particular we believe that the local injectivity radius around a typical vertex will go to infinity on such a surface as $N \to \infty$.

We would like to present here a natural quadrangulation that might describe such a local limit in the context of quadrangulations. A variant for triangulations might describe the limit with a specific supercritical random tree.

There exist nice and useful bijections between maps and labeled trees especially the so-called Schaeffer bijection. A variant of the UIPT (for quadrangulation) can be constructed from a labeled critical Galton-Watson tree conditioned to survive, see [CMM12] for details. Here we propose the study of a random quadrangulation constructed from a labeled super critical Galton-Watson trees.

Consider T_3 the full ternary tree given with a root vertex $\rho \in T_3$ and embedded in the plane $\mathbb{R}^2$. Assign independently to each edge e of the tree a random variable d_e uniformly distributed over $\{-1, 0, +1\}$. This procedure yields a labeling ℓ of the tree T_3 by setting the label of any vertex u as the sum of the d_e's along the geodesic line between ρ and u.

A *corner c* of the tree T_3 is an angular sector between two adjacent edges. There is a natural (partial) order on the corners of T_3 given by the clockwise contour of the tree T_3. We then extend the Schaeffer construction to the labeled tree (T_3, ℓ) as follows: For each corner c of T_3 associated to a vertex of label l, draw an edge between c and the first corner in the clockwise order whose associated vertex has label $l - 1$. Consider the quadrangulation obtained using only the edges added and not the original tree we started with. T_3 can be replaced by any tree.

It can be checked that all these edges can be drawn such that they are non-crossing and the resulting map is a infinite quadrangulation (with a root vertex ρ) that we call the Stochastic Hyperbolic Infinite Quadrangulation. It should be thought as a hyperbolic analogue of the UIPT/Q.

Here are several questions and observations regarding SHIQ:

- Does the SHIQ admits spatial Markovity? If it is indeed a local limit then yes.
- Starting with a super critical Galton Watson tree it easily follows that a.s. the quadrangulation has exponential volume growth. Estimate it. Are there limit theorem for ball size analogous to the branching process theory?
- Does the SHIQ has positive anchored expansion a.s. (see [Vir00] for the study of anchored expansion). This will imply positive speed and bounds on return probability.
- Using [BLS99] it is possible to show that simple random walk has positive speed.
- Is the sphere at infinity topologically S^1? Does SRW converges to a point on the sphere at infinity? Is the sphere at infinity the Martin boundary? See [Anc88] for details.
- Show that the Self Avoiding Walk is a.s. ballistic on the SHIQ? Adapt the theory of Poisson Voronoi percolation on the hyperbolic plain [BS01a] to the SHIQ. Study SHIQ coupled with spin systems such as Ising as for the UIPQ.

6.4   Sphere Packing of Graphs in Euclidean Space

One way to extend the notion of planar graphs in order to hopefully make initial steps in the context of three dimensional random geometry is to consider graphs sphere pack in $\mathbb{R}^3$. Some partial results extending ideas from planar circle packing to higher dimension were presented in [MTTV98, BS09] and [BC11]. The general theory of packing was recently developed by Pierre Pansu in [Pan]. See [BC11] for a collection of problems on the subject.

Maybe an extension of Schaeffer's bijection can used to create graphs sphere packed in $\mathbb{R}^3$. In Schaeffer's bijection the edges of a planar tree are labeled $-1, +1$ or 0. Walking around the tree as in depth first search and summing the labels, this defines a height function on the vertices, two values for each vertex. If an edge is added between any vertex and the closest vertex in the direction of the walk with a smaller height a quadrangulation is generated.

We hope that replacing the tree by a planar graph in a related recipe will create a packable graph. Let G be a planar graph, $f : G \to \mathbb{Z}$, with value differ by at most one between neighbors. Circle pack G in the Euclidean plain, for any vertex $v \in G$, add an edge from v to the vertex u which is among the closest to v in the Euclidean metric, with $f(u) < f(v)$, (were vertices are identified with the center of the corresponding circles).

Open problem 6.8. *Is the resulting graph a sphere packed in $\mathbb{R}^3$?*

Start with G the square grid. By [MTTV98] we know that packable graphs has separation function bounded by $n^{d-1/d}$. Can this be used to construct a counter example by maybe realizing large expanders in this way?

If the conjecture is true than a natural family of packable graph (perhaps) can be obtained by taking G to be a random quadrangulation and f the Gaussian free field on it. We don't know an example of a transient graph which does not contain a transient subgraph which is sphere packed in $\mathbb{R}^3$.

Theorem 6.9. *Assume G is a finite vertex transitive graph which is sphere packed in $\mathbb{R}^d$. The diameter of G is bigger than $C_d|G|^{1/d}$. For some universal constant depending only on d.*

Exercise 6.10. (Level 3) Prove this by combining the fact from [MTTV98] that packable graphs has separation function bounded by $n^{d-1/d}$ and Theorem 2.1.

For planar finite vertex transitive graphs this follows also from a known structure theorem [FI79].

Chapter 7
Growth and Isoperimetric Profile of Planar Graphs

In this section we review a joint work with Panos Papasoglu, see [BP11], in which the following is proved:

Definition 7.1. Let Γ be a locally finite graph. If v is a vertex of Γ we denote by $B(v, n)$ the ball of radius n centered at v. For a graph Γ we denote by $|\Gamma|$ the number of vertices of Γ. Define the *growth function* of Γ at a by

$$V(a, n) = |B(a, n)|$$

Theorem 7.2. *Let Γ be a planar graph such that the volume function of Γ satisfies $V(2n) \leq C V(n)$ for some constant $C > 0$. Then for every vertex v of Γ and $n \in \mathbb{N}$, there is a domain Ω such that*

1. *$B(v, n) \subset \Omega$,*
2. *$\partial\Omega \subset B(v, 6n)$,*
3. *$|\partial\Omega| \leq C \cdot n$.*

Definition 7.3. A graph Γ is said to be *doubling* if there is a constant $C > 0$ such that for all $a, b \in \Gamma$ and $n \in \mathbb{N}$, $V(a, 2n) \leq C V(b, n)$. We say then that C is a *doubling constant* for Γ.

If Γ is a doubling graph then the degree of vertices is uniformly bounded. Note that for any $d \geq 1$, there are planar graphs with the doubling property such that for any $v \in \Gamma$ and $n \in \mathbb{N}$, $V(a, n)$ is of order n^d, see e.g. the last section of [BS01b].

We say that a graph Γ corresponds to a *tessellation* of $\mathbb{R}^2$ if there is a $k \in \mathbb{N}$ such that all components of $\mathbb{R}^2 - \Gamma$ are bounded regions with at most k sides.

Definition 7.4. Let (X, d) be a metric space. An ϵ-*net* N of X is a set such that $d(v_1, v_2) > \epsilon$ for all $v_1, v_2 \in N$ and N is maximal set with this property.

We remark that if N is an ϵ-net of X then X is contained in the ϵ-neighborhood of N.

I. Benjamini, *Coarse 3Geometry and Randomness*, Lecture Notes in Mathematics 2100,   59
DOI 10.1007/978-3-319-02576-6_7,
© Springer International Publishing Switzerland 2013

The main result is the following:

Theorem 7.5. *Let Γ be a doubling planar graph. Then there is a constant α so that for every vertex $v \in \Gamma$ and $n \in \mathbb{N}$ there is a finite domain Ω such that $B(v,n) \subset \Omega$, $\partial\Omega \subset B(v,6n)$ and $|\partial\Omega| \leq \alpha n$.*

Krikun [Kri04] has shown a similar theorem for the uniform infinite planar triangulation (UIPT) introduced in [AS03]. The volume doubling property does not hold for the UIPT, still an asymptotic version should hold: for any vertex v, for large enough n, $B(v,2n)$ contains order 1 disjoint balls of radius $n/2$ a.s. and the proof below will adapt to give Krikun's result. The asymptotic volume growth of balls in the UIPT is order n^4, up to polylog's, see [Ang03], thus a weaker result with a polylog correction follows from our result.

This shows that the volume and the isoperimetric profile function are related for planar graphs. Recall the definition of the isoperimetric profile function of a graph (see Chap. 1):

Definition 7.6. Let Γ be a locally finite graph and let $V(n)$ be the volume function of Γ. Then the *isoperimetric profile* function of Γ, $I_\Gamma : \mathbb{N} \to \mathbb{N}$ is defined by:

$$I_\Gamma(n) = \inf_\Omega\{|\partial\Omega| : \Omega \subset \Gamma, |\Omega| \leq n\},$$

where Ω ranges over all subgraphs of Γ.

From the result above we obtain the following:

Corollary 7.7. *Let Γ be a doubling planar graph with volume function $V(n)$ and isoperimetric profile function $I_\Gamma(n)$. Let $\varphi(n) = \inf\{k : V(k) \geq n\}$. Then there is a constant α such that*

$$I_\Gamma(n) \leq \alpha\varphi(n), \quad \forall n \in \mathbb{N}$$

Proof Sketch of Theorem 7.5. Let v be any vertex of Γ. Consider the balls $B(v,n)$, $B(v,3n)$. Let N be an n-net of $\partial B(v,3n)$. For each vertex w of N consider $B(w,n/2)$. Note that all such balls are disjoint since N is an n-net. Also all these balls are contained in $B(v,4n)$. So, by the doubling property, we can have only boundedly many such balls, that is $|N| \leq \beta$, where β does not depend on n. Consider now the balls $B(w,2n)$ for all $w \in N$. $\partial B(v,3n)$ is contained in the union of these balls. Construct a closed curve that "blocks" v from infinity as follows: if $w_1, w_2 \in N$ are such that $d(w_1,w_2) \leq 2n$ then we join them by a geodesic. So replace $\partial B(v,3n)$ by the "polygonal line" that we define using vertices in N. This "polygonal line" blocks v from infinity and has length at most $2n\beta$. There are some technical issues to take care of, for example $\partial B(v,3n)$ might not be connected (and could even have "large gaps") and the geodesic segments have to be chosen carefully. In particular the constants obtained will be slightly different from the ones in this sketch.   □

Some further comments.

Definition 7.8. A graph G admits unform volume growth $f(n)$, if there are $0 < c < C < \infty$, so that for all n, any ball of radius n in G satisfies,

$$cf(n) < |B(v,n)| < Cf(n).$$

For planar graph admitting arbitrarily large uniform polynomial growth, see Example 1.94 and [BS01b]. It is conjectured that planar graphs of uniform polynomial growth are recurrent for the simple random walk. It is also conjectured [AS03] that the UIPT is recurrent, see also [BS01b]. By the Nash-Williams sufficient condition for recurrence, it is enough to find infinitely many disjoint cutsets $\{C_i\}$ separating the root from infinity. so that $\sum |C_i|^{-1} = \infty$. The theorem above is a step in that direction, still we don't know if planar graphs of uniform polynomial growth admits such cutsets? Maybe not.

Assume G is a planar graph or triangulation of uniform polynomial growth n^d, $d > 2$, by the theorem above G admits bottlenecks. This suggests that the t simple random walk on G will be subdiffusive, as it will spend a lot of time in domains with small boundary before exiting. That is,

Conjecture 7.9. The expected distance to the root by time t is bounded by t^α for $\alpha < 1/2$. Does $\alpha = d^{-1}$?

What about a high dimensional generalization? A d-sphere packing is a collection of d-dimensional balls with disjoint interiors. Associated to the packing an unoriented graph $G = (V, E)$ called the *d-tangency graph*, where vertices corresponds to the d-balls and edges are between any two tangent balls, see [BC11]. Is it the case that for any d a d-tangency graph with the doubling property admits cutsets outside a ball of radius n of size n^{d-1}?

Let G be a planar triangulation which is doubling and further assume all balls has growth r^d, $d > 2$ up to a multiplicative constant. Is there such G for which all complements of balls are connected, for all balls? Or as in the UIPT, the complements of some balls admit several connected components, some of size proportional to the ball?

Open problem 7.10. *Is $p_c < 1$ for planar triangulation of uniform growth faster than r^2. If this holds perhaps it is true that $p_c = \frac{1}{2}$.*

Chapter 8
Critical Percolation on Non-Amenable Groups

8.1 Does Percolation Occurs at the Critical Value?

For a given graph G, let $\theta_G(p) = P_p(0 \leftrightarrow \infty)$ (or just $\theta(p)$ when G is clear from the context). From the definition of p_c we know that $\theta(p) = 0$ for any $p < p_c$, and $\theta(p) > 0$ whenever $p > p_c$. A major and natural question that arises is: Does $\theta(p_c) = 0$ or $\theta(p_c) > 0$?. In this section we show that the answer to this question depends on the graph and prove that $\theta(p_c) = 0$ for a large variety of regular graphs.

A known conjecture regarding transitive graphs is the following:

Conjecture 8.1. If G is vertex transitive then $\theta(p_c) = 0$.

The following are examples of graphs for which there is no percolation at the critical value.

Example 8.2.

1. $\theta_{\mathbb{Z}^2}(p_c) = 0$. See [Gri99] for a proof.
2. For sufficiently large $d \in \mathbb{N}$, $\theta_{\mathbb{Z}^d}(p_c) = 0$. Currently this is known for $d \geq 19$ [HS90], the proof is thought to be extendable to any $d \geq 6$.

One of the biggest open questions in probability is to show that $\theta_{\mathbb{Z}^3}(p_c) = 0$.

Open problem 8.3. *Let $H = \mathbb{Z}^2 \times \mathbb{Z}_2$ be the graph obtained by taking two copies of $\mathbb{Z}^2$ and connecting each of the corresponding vertices $v \in \mathbb{Z}^2$ by an edge. We call this graph the Sandwich graph. Show that $\theta_H(p_c) = 0$.*

Exercise 8.4. (Level 1) One question that can be answered regarding the Sandwich graph is the following: Show $p_c(H) < p_c(\mathbb{Z}^2)$.

The following open problem might be helpful in proving that $\theta_{\text{sandwich graph}}(p_c) = 0$.

Open problem 8.5. *Show that if a finite energy, invariant percolation on $\mathbb{Z}^2$ which satisfy the FKG inequality, percolates a.s. than it also percolates in the half plans*

I. Benjamini, *Coarse 3Geometry and Randomness*, Lecture Notes in Mathematics 2100,   63
DOI 10.1007/978-3-319-02576-6_8,
© Springer International Publishing Switzerland 2013

almost surely. See [HM09] for background on the problem and an example that shows that the FKG inequality is necessary.

In Sect. 8.3 we will review the paper [BLPS99b] and show that one can say a lot about percolation in the critical value in nonamenable graphs, i.e. $h(G) > 0$. In general, $h(G) > 0$ does not imply that $\theta_G(p_c) = 0$, however, we will show that for Cayley graphs this implication does hold (See Theorem 8.18).

We see that in many cases, the percolation at the critical value dies out (or is expected to die out), meaning that there is no infinite cluster when $p = p_c$. This phenomenon is called second-order phase transition. However, there are also many examples of graphs where $p_c < 1$ yet $\theta(p_c) > 0$. In this case the phenomenon is called first-order phase transition.

Exercise 8.6. (Level 3) Which of the phenomenons occur in the tree in which the vertices at level n^2 have three offsprings whereas all the other vertices have two offsprings?

Let us present another simple phenomenon exhibiting a first order phase transition:

Exercise 8.7 ([Dek91]). (Level 4) Let T be the tree generated by a Galton-Watson process with offspring distribution $\mu_1 = p$ and $\mu_3 = 1 - p$. We denote the associated probability measure by $\mathbb{P}_p$, and consider the event

$$B = \left\{ \begin{array}{l} \text{There exists a complete binary tree starting} \\ \text{from the root embedded in the random tree } T \end{array} \right\}.$$

Show there exists $p_c \in (0, 1)$ such that if $p < p_c$ then $\mathbb{P}_p(B) = 0$ whereas when $p \geq p_c$ we have $\mathbb{P}_p(B) > 0$.

Extending Dekking's result from the exercise above to percolation on other graphs seems hard. For example, denote by H_k a k-regular planar triangulation.

Open problem 8.8. *Show that at the critical probability, for the percolation clusters on H_k to a.s. contain a full binary tree, this probability is positive.*

8.2   Invariant Percolation

In Sect. 4.1 we defined Bernoulli bond percolation. One can define more general bond percolation processes on a graph. Namely, given a graph $G = (V, E)$, any distribution on $\{0, 1\}^E$ is a bond percolation process. Among the family of percolation processes we distinguish percolation which are invariant under the group of automorphisms of the graph. More formally we define:

Definition 8.9. Invariant percolation on a graph $G = (V, E)$ is a distribution on $\{0, 1\}^E$ that is invariant under all automorphisms of G. Sometimes, we will refer to invariant percolation as a random subset of E, whose law is invariant under the automorphisms of G. These definitions are of course equivalent, and it is a matter of taste and notation which one to use. When one uses the notation $\{0, 1\}^E$ an edge $e \in E$ is said to be open in $\omega \in \{0, 1\}^E$ if $\omega(e) = 1$.

Here is a first example for invariant percolation.

Example 8.10. Let $G = \mathbb{Z}$. Denote $\{\omega_1, \omega_2\} \subset \{0, 1\}^{E(\mathbb{Z})}$, where $\omega_i = \{\{n, n + 1\}\}_{n=i \bmod 2}$. Clearly, neither ω_1 or ω_2 are invariant percolation. However, if we let $\omega = \omega_1$ with probability $\frac{1}{2}$, and $\omega = \omega_2$ with probability $\frac{1}{2}$, then it is easy to verify that the law of ω is invariant under the automorphisms of $\mathbb{Z}$, and hence ω is an invariant percolation.

Let us start the discussion on invariant percolation with a little warmup exercise. We saw in Theorem 4.1 that the critical parameter $p_c(\mathbb{Z}^2)$ for Bernoulli percolation on $\mathbb{Z}^2$ is between $\frac{1}{4}$ and $\frac{3}{4}$. Assume μ is an invariant percolation on $\mathbb{Z}^2$ such that for any fixed edge, $\mathbb{P}_\mu(e$ is open $) > 0.9 > p_c(\mathbb{Z}^2)$. Does it imply that

$$\mathbb{P}_\mu(\text{there exists an infinite open cluster}) > 0?$$

Answer. No. consider the following example: Choose a large k and observe the following tiling of the space. Denote $S_k = [0, k]^2 \cap \mathbb{Z}^2$. We define E_k to be the set of edges both their endpoints belong to S_k. Taking translations of S_k and E_k by $\mathbb{Z}^2 \times (k+1, k+1)$ gives a tiling of the space with no infinite component. We denote the set of open edges by ω_k. Consider the following distribution: Choose uniformly a translation from $\{(i, j)\}_{0 \le i, j \le k}$ and apply it to ω_k. This gives an invariant percolation satisfying $\mathbb{P}_\mu(e$ is open$) > 0.9$ whenever k is large enough, but clearly

$$\mathbb{P}_\mu(\text{there exists an infinite open cluster}) = 0.$$

Here is an important exercise.

Exercise 8.11. (Level 3) Is it possible to make a similar construction on the 3 regular tree?

For a finite graph K, set

$$\alpha_K = \frac{1}{|V(K)|} \sum_{x \in V(K)} \deg_K(x),$$

where $\deg_K(x)$ refers to the degree of x in the graph K. Given a graph G let

$$\alpha(G) = \sup\{\alpha_K; K \subseteq G \text{ is a finite subgraph}\}.$$

Exercise 8.12. (Level 1) Show that if G is a d-regular graph then

$$\alpha(G) + h(G) = d,$$

where $h(G)$ is the Cheeger constant (see Definition 1.2)

The following very important theorem is the main result of this subsection.

Theorem 8.13. *Let G be a Cayley graph and let ξ be an invariant percolation on G. If $\mathbb{E}[deg_\xi(e)] > \alpha(G)$ then ξ has an infinite clusters with positive probability.*

Before turning to the proof we will prove a very useful lemma, known as the Mass-Transport-Principle. This principle is intensively used in the following. Also see Sect. 5.3 for generalizations.

Lemma 8.14 (The Mass-Transport-Principle). *Let G be a countable group. If $M : G \times G \rightarrow [0, \infty]$ satisfy $f(\gamma v, \gamma w) = f(v, w)$ for every $\gamma, v, w \in G$, then*

$$\sum_{w \in G} M(v, w) = \sum_{w \in G} M(w, v), \quad \forall v \in G.$$

Proof. The proof just uses the basic structure of the group G or equivalently of its Cayley graph.

$$\sum_{w \in G} M(v, w) = \sum_{\gamma \in G} M(v, v\gamma) = \sum_{\gamma \in G} M(e, \gamma) = \sum_{\gamma \in G} M(e, \gamma^{-1})$$

$$= \sum_{\gamma \in G} M(\gamma, e) = \sum_{\gamma \in G} M(v\gamma, v) = \sum_{w \in G} M(w, v).$$

□

We can now turn to prove the main result of this section.

Proof of Theorem 8.13. For a vertex $v \in V$ denote by $K(v)$ its connected component in the percolation ξ. For every realization ξ of the invariant percolation and $v, w \in V$, we define the function $m(v, w, \xi)$ by

$$m(v, w, \xi) = \begin{cases} \frac{deg_\xi(v)}{|K(v)|}, & \text{if } K(v) \text{ is finite and } w \in K(v) \\ 0, & \text{otherwise.} \end{cases}$$

Note that if $K(v)$ is finite then $\frac{1}{|V(K)|} \sum_{x \in V(K)} deg_K(x) \leq \alpha(G)$. Let $M(v, w) = \mathbb{E}[m(v, w, \xi)]$. We think about $m(v, w, \xi)$ as the amount of mass transported from v to w in ξ. With this interpretation in mind, $M(v, w)$ is the expected amount of mass transported from v to w. Note that since G is a Cayley graph it is in particular transitive. Consequently, using the invariance of the percolation it follows that $M(v, w) = M(\gamma v, \gamma w)$ for every choice of $\gamma, v, w \in G$. Thus by the Mass transport principle (Lemma 8.14) $\sum_{w \in G} M(v, w) = \sum_{w \in G} M(w, v)$ for every $v \in G$. Next, we turn to estimate the last sum with $v = e$ in two different ways. First,

$$\sum_{w \in G} M(e, w) = \mathbb{E}\left[\sum_{w \in G} \frac{\deg_\xi(e)}{|K(e)|} \mathbb{1}_{w \in K(e)} \mathbb{1}_{|K(e)| < \infty}\right] \tag{8.1}$$

$$= \mathbb{E}\left[\deg_K(e) \mid |K(e)| < \infty\right] \cdot P(|K(e)| < \infty).$$

On the other hand

$$\sum_{w \in G} M(w, e) = \mathbb{E}\left[\sum_{w \in G} \frac{\deg_\xi(w)}{|K(w)|} \mathbb{1}_{e \in K(w)} \mathbb{1}_{|K(w)| < \infty}\right]$$

$$= \mathbb{E}\left[\sum_{w \in G} \frac{\deg_\xi(w)}{|K(e)|} \mathbb{1}_{w \in K(e)} \mathbb{1}_{|K(e)| < \infty}\right] \tag{8.2}$$

$$= \mathbb{E}[\alpha_K \mid |K(e)| < \infty] \cdot P(|K(e)| < \infty)$$

$$\leq \alpha(G) \cdot P(|K(e)| < \infty).$$

Thus if $P(|K(o)| < \infty) = 1$, we obtain that

$$\mathbb{E}\left[\deg_K(e)\right] \leq \alpha(G)$$

which contradicts the assumption of the Theorem and therefore completes the proof.
□

With Olle and Oded we long ago asked,

Open problem 8.15. *Is there an invariant finite energy percolation X on $\mathbb{Z}^d$, which a.s. percolates and satisfies $p_c(X) = 1$?*

Exercise 8.16. (Level 3) Partition $\mathbb{Z}$ into infinitely many infinite indistinguishable sets. Here, indistinguishable is in the sense that they have any invariant property with the same $0 - 1$ probability.

Exercise 8.17. (Level 3) Show that if the pair correlation for neighboring vertices, in an invariant site percolation on the 3-regular tree, are sufficiently close to 1, then there are infinite clusters, of either the open or the closed vertices. Construct such invariant percolations, in which both closed and open vertices percolate.

8.3 $\theta(p_c) = 0$ When $h(G) > 0$

This section is devoted to the main theorem of this chapter, which originally appeared in [BLPS99b], regarding percolation in non amenable Cayley graph at the critical value.

Theorem 8.18. *If G is a Cayley graph with $h(G) > 0$, then $\theta_G(p_c) = 0$.*

Since the proof is quite long we only give a sketch of it. For a full detailed version see [BLPS99b].

Proof sketch 8.19. *Assume $\theta_G(p_c) > 0$ and denote the Bernoulli percolation with parameter p_c by ω_{p_c}. There are two cases to consider[1]:*

- *There is almost surely a unique infinite cluster in ω_{p_c}*
- *There are almost surely infinitely many infinite clusters in ω_{p_c}.*

We begin with the first case. Denote the unique infinite cluster of ω_{p_c} by U. For every $x \in V_G$, let $U(x) \subseteq V_G$ be the set of closest neighbors of x in U. We define a new invariant percolation on G denoted γ_ϵ as follows. Let ξ_ϵ be ω_{p_c} intersected with a $(1 - \epsilon)$ Bernoulli bond percolation, which is independent of ω_{p_c}. Then ξ_ϵ is a Bernoulli bond percolation with parameter $(1 - \epsilon)p_c$. Next we declare an edge $e = (x, y) \in E_G$ to be open in γ_ϵ if the following conditions hold

1. *$d_G(x, U(x)) \leq \frac{1}{\epsilon}$,*
2. *$d_G(y, U(y)) \leq \frac{1}{\epsilon}$,*
3. *$U(x)$ and $U(y)$ are in the same connected component of ξ_ϵ.*

Note that $\lim_{\epsilon \downarrow 0} \mathbb{E}[e \in \gamma_\epsilon] \to 1$. Thus Theorem 8.13 implies that γ_ϵ contains infinite clusters with positive probability for every small enough $\epsilon > 0$. It is not hard to see that if γ_ϵ contains infinite clusters, then so does ξ_ϵ. This however contradicts the fact that ξ_ϵ is a Bernoulli percolation with parameter $(1 - \epsilon)p_c < p_c$.

We now turn deal with the second case. Again we emphasize that more details can be found in [BLPS99b]. Assume $\theta_G(p_c) > 0$ and that there are infinitely many infinite clusters in ω_{p_c} with probability 1. We say that a vertex v is an encounter point if it belongs to an infinite cluster, but the removal of v splits the cluster into at least three infinite clusters. One can show that if there are infinitely many infinite clusters, then there are a.s. infinitely many encounter points. Next define a random graph Γ as follows. Let the vertex set of Γ be the set of encounter points, and add an edge between two encounter points u and v if v is the closest encounter point to u in the connected component of u. One can now show that Γ is in fact a forest (i.e. doesn't contain cycles), which consists of infinitely many infinite trees. In addition, one can show that if v is an encounter point, then there is at least one encounter point in each of the infinite clusters that are created by removing v. Therefore, the degree of any vertex in Γ is at least 3. Let γ_ϵ be a Bernoulli percolation with parameter ϵ on G. Define the subforest $\Gamma_\epsilon \subset \Gamma$ by letting the vertex set be $V(\Gamma_\epsilon) = V(\Gamma)$ and declare an edge $\{u, v\}$ in $E(\Gamma)$ open in $E(\Gamma_\epsilon)$ if u and v belong to the same component of $\omega_p - \gamma_\epsilon$. One can now complete the proof by showing that for $\epsilon > 0$ sufficiently small γ_ϵ contains unbounded connected components. Since this implies that $\omega_{p_c} - \gamma_\epsilon$ also contains infinite clusters, we get a contradiction to the definition of $p_c(G)$.

---

[1]The reason we need to deal only with these cases is explained in Chap. 4

Chapter 9
Uniqueness of the Infinite Percolation Cluster

Since we know that for $p > p_c$ there is an infinite percolation cluster a.s., it is natural to ask whether it is unique or not. As it turns out the answer to this question leads to a rather rich landscape with applications in group theory and many still open problems. In this section we study the question of the number of infinite clusters in percolation configurations in the regime $p > p_c$.

For completely general graphs, there is very little we can say. Here are some examples for possible situations:

1. Attach two copies of $\mathbb{Z}^2$ by an edge, then for $p > 1/2$ there are either one or two infinite clusters.
2. Attach an infinite binary tree to the origin of $\mathbb{Z}^3$, then for $p \in \left(p_c(\mathbb{Z}^3), 1/2\right]$ there is a unique infinite cluster, while for $p \in (1/2, 1)$ there are infinitely many.

Other such combinations of graphs can lead to fairly complex dependence of the number of clusters on p.

9.1 Uniqueness in $\mathbb{Z}^d$

Uniqueness of the infinite percolation cluster on $\mathbb{Z}^d$ (and in fact, on any amenable transitive graph) is one of the classical results of percolation theory. Before describing Burton and Keane's argument for uniqueness of the infinite cluster in $\mathbb{Z}^d$ we first state a general result about any transitive graph.

Theorem 9.1. *In Bernoulli percolation on any transitive graph, the number of infinite clusters is an a.s. constant. In addition it can achieve exactly one of three values: 0, 1, and* ∞.

The fact that the number of infinite clusters is constant a.s. follows from ergodicity of the percolation measure w.r.t translations and the fact that the number of infinite cluster is a measurable and invariant under translations random variable.

I. Benjamini, *Coarse 3Geometry and Randomness*, Lecture Notes in Mathematics 2100, DOI 10.1007/978-3-319-02576-6_9,
© Springer International Publishing Switzerland 2013

The fact that it cannot be any finite number greater than 1, is proved using a finite energy argument similar to the one in the proof below. See [Gri99] for more details.

Theorem 9.2. *Consider Bernoulli edge percolation with parameter p on $\mathbb{Z}^d$. If $\theta(p) > 0$, then there exists a unique infinite cluster a.s.*

In fact, the proof works in the much more general class of transitive amenable graphs:

Theorem 9.3. *Consider Bernoulli edge percolation with parameter p on a transitive amenable graph G. Then the number of infinite clusters is either 0 or 1.*

Proof. Due to Theorem 9.1, it is enough to rule the case of infinitely many infinite clusters, so suppose towards contradiction that there are infinitely many infinite clusters. A vertex $v \in V$ is said to be a *trifurcation* point if

  (i) v belongs to an infinite cluster.
 (ii) There exist exactly three open edges incident to v.
(iii) Deleting v and its three incident open edges splits its infinite cluster into three disjoint infinite clusters.

Fix some v, and denote by p_t be the probability that it is a trifurcation point. By transitivity, p_t is independent of v. Since there are almost surely infinitely many infinite clusters, for large enough R there is a positive probability that $B(v, R)$ intersects at least three infinite clusters. Thus with positive probability there are three infinite clusters in the complement of $B(v, R) =$ that reach the boundary $\partial B(v, R)$. Note that this event depends only on edges outside $B(v, R)$ and since there are only finitely many edges inside $B(v, R)$ changing their value will still give an event with positive probability. By changing the value of all edges inside the box, so that the only open edges in $B(v, R)$ are three disjoint paths connecting the three infinite clusters to a single vertex $u \in B(v, r)$, we ensure that u has positive probability of being a trifurcation point. Thus $p_t > 0$.

Next we get a contradiction by showing that for every vertex transitive amenable graph $p_t = 0$. Consider some finite set $W \subset G$, and let T_W be the set of trifurcation points in W. Call a point $v \in T_W$ an *outer* point if at least two of the disjoint infinite clusters one achieves by removing v don't contain points from T_W. If there are trifurcation points in the graph one can choose W that includes at least one trifurcation point. Since W is finite it follows that T_W must contain at least one outer point. We will now show by induction on $|T_W|$ that the removal of all trifurcation points in T_W will result in at least $|T_W| + 2$ disjoint infinite clusters intersecting W. In the case $|T_W| = 1$ this follows directly from the definition of outer trifurcation points. Next, assume it holds for $|T_W| = j$ and suppose T_W is a set of $j + 1$ trifurcation points. Let v be an outer member of T_W. According the induction assumption the removal of all vertices in $T_W \setminus \{v\}$ splits the infinite cluster into at least $j + 2$ disjoint ones. Since v is an outer point the removal of it gives one more infinite cluster (two if it is not connected to any other point in A), completing the induction proof.

Finally, denote by $T(W)$ the number of trifurcations in W, i.e. $T(W) = |T_W|$. The last induction implies that $T(W) \leq |\partial W| - 2$. Using the transitivity of G it follows that $\mathbb{E}[T(W)] = p_t|W|$, and thus

$$p_t \leq \frac{|\partial W| - 2}{|W|}.$$

Since G is amenable, W can be chosen so that the right hand side of the above inequality becomes arbitrarily small. Thus $p_t = 0$. $\square$

Exercise 9.4. (Level 3) Show that after removing from $\mathbb{Z}^3$ the range of a simple random walk, an infinite connected component is left a.s.

Exercise 9.5. (Level 3) Prove uniqueness of the infinite cluster for half space and quarter space.

Open problem 9.6. *Prove uniqueness of the infinite cluster for bounded degree graphs which are rough isometric to $\mathbb{Z}^d$?*

9.2 The Uniqueness Threshold

Since in transitive graphs when infinite clusters exists their number is either 1 or ∞, the following definition makes sense.

Definition 9.7. The *uniqueness threshold* of a connected graph G is defined by

$$p_u = p_u(G) = \inf\{p \geq p_c : \mathbb{P}_p \text{ (there exists a unique infinite cluster)} = 1\}.$$

Note that from the definition we have

$$0 \leq p_c \leq p_u \leq 1 \tag{9.1}$$

Example 9.8. It is easy to verify that in Bernoulli percolation on a binary tree, the number of infinite clusters is 1 only when $p = 1$. Hence

$$p_u \text{ (binary tree)} = 1.$$

Example 9.9. From Theorem 9.2, it follows that $p_u(\mathbb{Z}^d) = p_c(\mathbb{Z}^d)$.

In [BS96b] it was conjectured that the converse of Theorem 9.3 is also true:

Conjecture 9.10. For a vertex transitive graph G, $p_c(G) < p_u(G)$ if and only if $h(G) > 0$.

As stated before, one direction of Conjecture 9.10 is established in Theorem 9.3. The other direction, i.e. showing that $p_c < p_u$ whenever $h(G) > 0$ has turned out to be a more difficult problem. Partial progress has been made in number of cases and directions, see [Lyo09] for a review of the partial results. One such progress was made by Pak and Smirnova-Nagnibeda, see [PSN00]. Using a criterion we will discuss in the next section and an inequality of Mohar (see [Moh88]) they proved that any nonamenable Cayley graph admits a generating set for which $p_c < p_u$. As it turns out the last result is useful outside of probability, see e.g. [Hou11]. In fact in [PSN00] Pak and Smirnova-Nagnibeda also proved that in order to prove the conjecture above it is sufficient to prove the following one:

Conjecture 9.11. In the context of bounded degree graphs the property $p_c < p_u$ is a rough isometric invariant.

It is possible to naturally define a *finitary* analogue of the intermediate phase $p_c < p < p_u$ of percolation on infinite graphs. Given a family of finite vertex transitive graphs, G_n, and p, consider a neighboring pair of vertices in each G_n, conditioned to be in the same p-percolation open component. Is their distance in the component of G_n a tight family of random variables? The p's for which it is not tight, is the analogous phase.

Exercise 9.12. (Level 3) Prove that $p_u(T_d \times \mathbb{Z}) \leq p_c(\mathbb{Z}^2)$, where T_d denotes the d-regular tree.

This leads to one family of non-trivial examples of graphs consistent with Conjecture 9.10:

Theorem 9.13 ([BS96b, GN90]). *If d is sufficiently large then $p_c(T_d \times \mathbb{Z}) < p_u(T_d \times \mathbb{Z})$.*

Schonmann proved in [Sch99] that Bernoulli percolation on $T_d \times \mathbb{Z}$ at p_u has infinitely many infinite clusters almost surely. We will see in Sect. 9.5 that this is not possible for planar graphs.

Ancona [Anc88] proved that on nonamenable hyperbolic graphs, the Green function of the simple random walk is quasi multiplicative. In the sense that if z is on a geodesic from x to y, then the probability SRW starting at x visits y, is proportional to the probability of visiting y, while passing via z.

Show quasi multiplicativity for percolation connectivity, at the critical percolation probability, or even for p slightly larger or possibly any p? That is,

Open problem 9.14. *For which p's, There is $C < \infty$, so that $P_p(x$ connected to $y) < C P_p(x$ connected to $z) P_p(z$ connected to $y)$, for any two vertices x, y and z on a geodesic between them.*

Proving this for some $p > p_c$ will imply $p_c < p_u$ for nonamenable hyperbolic graphs.

9.3 Spectral Radius

In this section we prove a simple criterion which guarantees that $p_c < p_u$.

Definition 9.15. Given a graph $G(V, E)$ we denote by $\{p_n(v, w)\}_{n \in \mathbb{N}, v, w \in V}$ the heat kernel of the random walk, where $p_n(v, w)$ is the probability that a Simple Random Walk (SRW) starting at v hit w at time n. In addition we define the spectral radius of G as

$$\rho(G) = \limsup_{n \to \infty} \sqrt[n]{p_n(v, v)}.$$

Note that the right hand side is independent of v whenever the graph is connected.

In the following exercises we investigate the behavior of $p_n(\cdot, \cdot)$ in trees.

Exercise 9.16. (Level 1) Show that in $\mathbb{Z}$ one has $p_{2n}(0, 0) = \binom{2n}{n} \cdot 2^{-2n} \sim \frac{c}{\sqrt{n}}$ and $p_{2n+1}(0, 0) = 0$.

Exercise 9.17. (Level 2) What is $p_{2n}(0, 0)$ for $G = T_d$?

Example 9.18. In $\mathbb{Z}^d$ the heat kernel satisfies $p_n(v, w) \leq \frac{c}{n^{\frac{d}{2}}}$ and therefore $\rho(\mathbb{Z}^d) = 1$.

Exercise 9.19. (Level 2) Show that $\rho(T_d) \xrightarrow{d \to \infty} 0$ and $\rho(T_d \times \mathbb{Z}) \xrightarrow{d \to \infty} 0$.

Exercise 9.20. (Level 4) Show that for every bounded degree graph, $h(G) > 0$ if and only if $\rho(G) < 1$. Hint: Use the equivalence of (1) and (3) in Theorem 7.3 in [Pet09].

Next we state a criterion that implies $p_u < p_c$.

Theorem 9.21. *Assume G is vertex transitive graph and denote the degree of the vertices by d_G; assume in addition that $p > p_c$ and that $\rho(G) \cdot d_G \cdot p < 1$. Then there are infinitely many infinite clusters $\mathbb{P}_p$ a.s.*

In order to prove the last theorem we will need the next definition and the lemma followed by it.

Definition 9.22. Define the following Branching Random Walk (BRW): At time 0, start at $v_0 \in V$ with one particle. Then given that in time n there are N_n particles in the vertices $\{x_i\}_{i=1}^{N_n}$, the configuration of the particles at time $n + 1$ is defined as follows: For every $1 \leq i \leq N_n$ the ith particle gives birth to a new particle in each of its neighbors with probability p independently. After performing this process, the particle x_i disappears.

Lemma 9.23. *The following inequality holds:*

$$\mathbb{P}_p \left(u \text{ is connected to } v \right) \leq \mathbb{P}_p \left(BRW \text{ starting in } u \text{ ever hits } v \right).$$

Exercise 9.24. (Level 3) Prove Lemma 9.23.

Assuming Lemma 9.23, we now turn to prove Theorem 9.21.

Proof. Since $p > p_c$ and G is transitive, we know that $\mathbb{P}_p [v \leftrightarrow \infty]$ is independent of v and positive. We denote its value by α. Take $u, v \in V$ for which $d(u, v)$ is sufficiently large (to be chosen later). If there is a unique infinite cluster then by the FKG inequality (See [Gri99])

$$\mathbb{P}_p (v \leftrightarrow u) \geq \mathbb{P}_p (u, v \text{ belong to the infinite cluster}) \geq \alpha^2 > 0$$

Thus by Lemma 9.23, we get

$$\mathbb{P}_p (\text{BRW starting in } u \text{ ever hits } v) \geq \alpha^2.$$

This implies

$$\mathbb{P}_p (\text{BRW starting at } v \text{ hits } v \text{ again at time bigger than } 2d(v, u)) \geq \alpha^4. \quad (9.2)$$

On the other hand we know that $p_{2d(u,v)}(u, v) \sim \rho(G)^{2d(v,u)}$ and the number of particles that were born up to time $2d(u, v)$ is $\sim (p \cdot d_G)^{2d(v,u)}$. Recalling the assumptions of the theorem and applying a union bound we get

$$\mathbb{P}_p \begin{pmatrix} \text{BRW starting at } v \text{ hits } v \text{ again} \\ \text{at time bigger than } 2d(v, u) \end{pmatrix} \sim (p \cdot d_G \cdot \rho(G))^{2d(v,u)} \xrightarrow[d(v,u) \to \infty]{} 0 \quad (9.3)$$

Contradicting (9.2) and therefore the assumption of a unique infinite cluster. Since this is a vertex transitive graph and we are in the regime $p > p_c$ this is the only case we needed to exclude. $\qquad\square$

Example 9.25. The graph $G = T_d \times \mathbb{Z}$ for d large enough satisfies the conditions of the last theorem. Thus for large enough d we have $p_c(T_d \times \mathbb{Z}) < p_u(T_d \times \mathbb{Z})$.

Exercise 9.26. (Level 4) Gantert and Müller showed in [GM06] that critical branching random walk on a non-amenable graph is transient. Assume that G is non amenable and has one end. Does the range of the critical BRW have one end? (See Chap. 6 for the definition of ends).

9.4 Uniqueness Monotonicity

This subsection is based on the material found in [HP99]. We wish to know for any given Cayley graph G and $p \geq p_c(G)$ whether there is a unique or infinite cluster or infinitely many. So far we saw that $p_c \leq p_u$ however it is not clear from the definition that if $p > p_u$ then there are $\mathbb{P}_p$ almost surely only one infinite component, i.e. it is not clear whether this event is even monotone in p. In this section we show that under certain assumptions on the graph uniqueness is monotone, namely that for all $p > p_u$ there is a.s. a unique infinite cluster. The central tool we use is the mass transport principle (see Lemma 8.14).

Consider Bernoulli edge percolation with parameter $p \in [0, 1]$ on a Cayley graph G. Denote by $N = N(p)$ the number of infinite clusters in the percolation, and recall that N is almost surely a constant (which might depend on p). Formulating the question above we get: Is there a $p > p_u$ for which $N \neq 1$. It is reasonable to believe that if there is a unique infinite cluster at level p, then it takes up a large portion of the graph, and therefore it should not be possible for new infinite clusters to be created at level $p_1 > p$. We will now prove that this intuition is indeed correct.

Theorem 9.27. *Consider Bernoulli percolation on a Cayley graph G, and suppose $0 < p_1 < p_2 \leq 1$. If $\mathbb{P}_{p_1}(N = 1) = 1$, then also $\mathbb{P}_{p_2}(N = 1) = 1$.*

In particular there is a unique infinite cluster P_p-a.s. for every $p > p_u$.

Remark 9.28. Note that the condition that G is a Cayley graph can actually be weakened to the condition that G is transitive [HPS99]. However, the proof we present here uses the mass transport method, which we only proved for groups.

Proof. Recall that for a graph G we denote by $d_G(u, v) = d(u, v)$ the graph-distance between the vertices u and v in G. Suppose $0 < p_1 < p_2 \leq 1$. Let X_1 be edge percolation with parameter p_1 and let X_2 be an edge percolation obtained by taking the union of X_1 with an independent edge percolation with parameter $\frac{p_2 - p_1}{1 - p_1}$. With this construction we have that $X_1 \subset X_2$ and X_2 is an edge percolation with parameter p_2. The assertion of the theorem will follow once we show that every infinite cluster in X_2 contains some infinite cluster of X_1 a.s. This is, in view of the fact that $X_1 \subset X_2$ equivalent to showing that every infinite cluster in X_2 intersects some infinite cluster in X_1.

For $i = 1, 2$, denote by $C_i(v)$ the component of X_i containing v and by S_i the set of all vertices which belong to some infinite cluster of X_i. Finally, for any $v \in V$ we denote

$$D_i(u) = d(u, S_i).$$

Since we want to show that every infinite cluster of X_2 intersects S_1, we consider the distance from an infinite cluster of X_2 to S_1, and in particular the vertices that realize that distance. For $u \in V$, let $F(u)$ be the event that u is in an infinite X_2-cluster, but not in an infinite X_1-cluster, and u minimizes the distance to S_1 among its cluster in X_2, i.e.

$$B(u) = \{u \in S_2 \setminus S_1\} \cap \{D_1(u) = \min_{v \in C_2(u)} D_1(v)\}.$$

If there is an infinite X_2-cluster that does not intersect any infinite X_1-cluster, than that cluster must contain at least one vertex u which is closest to S_1. For this vertex, the event $B(u)$ occurs. Thus the proof will be complete as soon as we establish $\mathbb{P}(B(u)) = 0$.

Let $M(u)$ be the number of vertices in $C_2(u)$ that achieve the minimal distance to S_1 and

$$B^f(u) = B(u) \cap \{M(u) < \infty\}, \quad B^\infty(u) = B(u) \cap \{M(u) = \infty\}.$$

Thus $B(u) = B^f(u) \cup B^\infty(u)$ and therefore it is enough to prove that $\mathbb{P}(B^f(u)) = \mathbb{P}(B^\infty(u)) = 0$.

These two claims are proved in very different manners. We start with the event $B^f(u)$. This event almost allows us to select a finite set of vertices in a translation invariant manner, which is clearly impossible, but this is not exactly the case since there may be infinitely many clusters, each with a finite set of points where the distance to S_1 is minimized. To show that $\mathbb{P}(B^f(u)) = 0$, we use a mass transport argument: Let $X = (X_1, X_2)$ and define

$$m(x, y, X) = \begin{cases} \frac{1}{M(x)} & y \in C_2(x), y \text{ is a vertex closest to } S_1, B^f(x) \text{ occur.} \\ 0 & \text{otherwise} \end{cases} \quad (9.4)$$

From the definition of m we have

$$\sum_{y \in V} m(x, y, X) \leq 1$$

for every $x \in V$ (More precisely it equals 1 if $B^f(x)$ occurs and 0 otherwise). On the other hand

$$\sum_x m(x, y, X) = \sum_x m(x, y, X)\mathbb{1}_{B^f(y)} + m(x, y, X)\mathbb{1}_{(B^f(y))^c}$$
$$= \infty \cdot \mathbb{1}_{B^f(y)} + \sum_x m(x, y, X)\mathbb{1}_{(B^f(y))^c} \quad (9.5)$$

Thus by the mass transport principle we must have $\mathbb{P}(B^f(u)) = 0$.

We now move on to deal with the event $B^\infty(u)$. Partition $B^\infty(u)$ into the events $B_k^\infty(u) = B^\infty(u) \cap \{D_1(u) = k\}$. Since this is a countable partition it is enough to show that $\mathbb{P}(B_k^\infty(u)) = 0$ for any fixed k. Condition first on the entire configuration X_1. Next, since S_1 is now known, condition also on all edges in X_2 not incident to vertices within distance $k - 1$ from S_1. With this conditioning, every edge whose state is not known yet has conditional probability at least $\frac{p_2 - p_1}{1 - p_1} > 0$ to be open in X_2. This includes all edges going from vertices at distance k from S_1 towards S_1.

On the event $B_k^\infty(u)$ we now see the entire infinite cluster $C_2(u)$ (since it only reaches distance k from S_1), and there are infinitely many possible disjoint paths of length at most k connecting $C_2(u)$ to S_1. Thus when revealing the remaining edges of X_2, with probability 1, some (infinitely many) of these paths will become open (here we use the fact the their length is uniformly bounded), and therefore the infinite

X_2 cluster of u will be connected to an infinite X_1 cluster. Thus $\mathbb{P}(B_k^\infty(u)) = 0$ for every k, and therefore $\mathbb{P}(B^\infty(u)) = 0$. This completes the proof. □

We will not review the results regarding which graphs admit $p_u < 1$.

9.5 Nonamenable Planar Graphs

We finish our discussion on the uniqueness of the infinite cluster with a proof of a special case of Conjecture 9.10 proved via some Theorems in [BS01a] which we now present. These theorems apply primarily to lattices in the hyperbolic plane. See the section on the hyperbolic plane (Sect. 3) for additional background.

We start by recalling a definition from graph theory:

Definition 9.29. A ray in an infinite graph is an (semi) infinite simple path. Two rays are said to be equivalent if there is a third ray (which is not necessarily different from either of the first two) that contains infinitely many of the vertices from both rays. This is an equivalence relation. The equivalence classes are called ends of the graph.

Theorem 9.30. *Let G be a transitive, nonamenable, planar graph with one end. Then $0 < p_c(G) < p_u(G) < 1$, for bond or site percolation on G.*

We also have some understanding of the behavior at p_u:

Theorem 9.31. *Let G be a transitive, nonamenable, planar graph with one end. Then percolation with parameter p_u on G has a unique infinite cluster a.s.*

The proofs appear in Sect. 9.5.3.
Recall the following theorem:

Theorem 9.32 ([BLPS99a]). *Let G be a nonamenable graph with a vertex-transitive unimodular automorphism group (see definition below). Then critical Bernoulli bond or site percolation on G has no infinite components $\mathbb{P}_{p_c(G)}$ a.s.*

Conjecture 9.33. We conjecture that transient (simple) branching random walk has infinitely many ends on any vertex transitive graph.

Before we give proofs to Theorems 9.30 and 9.31, we have to introduce some notations and preliminaries from [BS01a].

9.5.1 Preliminaries

Given a graph G, let Aut(G) denote its automorphisms group. Recall that G is called *transitive* if Aut(G) acts transitively on the vertices of G, i.e. for every $x, y \in G$ there exists $\varphi \in$ Aut(G) such that $\varphi(x) = y$. We say that G is *quasi-transitive* if $V(G)/\text{Aut}(G)$ is finite; that is, there are finitely many $\text{Aut}(G)$ orbits in $V(G)$. Finally a graph G is *unimodular* if $\text{Aut}(G)$ is a unimodular group (i.e. $Aut(G)$ is a locally compact topological group whose left-invariant Haar measure is also right-invariant. For more details on this definition and more see [Pon86]).

Cayley graphs are unimodular, and any graph such that $\text{Aut}(G)$ is discrete is unimodular. See [BLPS99a] for a further discussion of unimodularity and its relevance to percolation.

Some of the results may be stated in much greater generality, not just for Bernoulli (i.e. independent) percolation but for much more general invariant distributions on configurations of edges.

Let $X = \mathbb{R}^2$ or $X = \mathbb{H}^2$. We say that an embedded graph $G \subset X$ in X is *properly* embedded if every compact subset of X contains finitely many vertices of G and intersects finitely many edges. Suppose that G is an infinite connected graph with one end, properly embedded in X. Let $G^\dagger$ denote the dual graph of G. We assume that $G^\dagger$ is embedded in X in the standard way relative to G: that is, every vertex $v^\dagger$ of $G^\dagger$ lies in the corresponding face of G, and every edge $e \in E(G)$ intersects only the dual edge $e^\dagger \in E(G^\dagger)$, and only in one point. If ω is a subset of the edges $E(G)$, then $\omega^\dagger$ will denote the set

$$\omega^\dagger = \left\{ e^\dagger \ : \ e \notin \omega \right\}.$$

Given $p \in [0, 1]$ and a graph G, we often denote the percolation graph of Bernoulli(p) bond percolation on G by ω_p.

Proposition 9.34. *Let G be a transitive, properly embedded, nonamenable, planar graph with one end, and let Γ be the group of automorphism of G. Then:*

(a) Γ *is discrete (and hence unimodular).*
(b) G can be embedded as a graph G' in the hyperbolic plane $\mathbb{H}^2$ in such a way that the action of Γ on G' extends to an isometric action on $\mathbb{H}^2$. Moreover, the embedding can be chosen in such a way that the edges of G' are hyperbolic line segments.

A sketch of the proof of (b) appeared already in [Bab97].

Proof. It follows from [Wat70] that G is 3-vertex connected, that is, every finite nonempty set of vertices $\emptyset \neq V_0 \subset V(G)$, has at least 3 vertices in $V(G) \setminus V_0$. Therefore, by the extension of Imrich to Whitney's Theorem, see [Imr75], the embedding of G in the plane is topologically unique, in the sense that in any two embedding of G in the plane, the cyclic orientation of the edges going out of the vertices is either identical for all the vertices, or reversed for all the vertices. This implies that an automorphism of G that fixes a vertex and all its neighbors is the identity. Therefore $\text{Aut}(G)$ must be discrete. For a discrete group, the counting measure is the Haar measure, and is both left and right invariant. Hence $\text{Aut}(G)$ is unimodular. This proves part (a).

Think of G as embedded in the plane. Call a component of $S^2 \setminus G$ a *face* if its boundary consists of finitely many edges in G. In each face f put a new vertex v_f, and connect it by edges to the vertices on the boundary of f. If this is done appropriately, then the resulting graph $\hat{G}$ is still embedded in the plane. Note that $\hat{G}$ together with all its faces forms a triangulation T of a simply connected domain in S^2. To prove (b) it is enough to produce a triangulation T' of $\mathbb{H}^2$ isomorphic with T such that the elements of $\mathrm{Aut}(T')$ extend to isometries of $\mathbb{H}^2$ and the edges of T' are hyperbolic line segments. There are various ways to do that. One of them is with circle packing theory (see, for example, [Bab97].) □

9.5.2 The Number of Components

In this subsection we discuss the correspondence between the number of infinite components in a nonamenable transitive graph and the number of infinite components in its dual. We start with the following Theorem:

Theorem 9.35. *Let G be a transitive, nonamenable, planar graph with one end, and let ω be an invariant bond percolation on G. Let k be the number of infinite components of ω, and $k^\dagger$ be the number of infinite components of $\omega^\dagger$. Then with probability one*

$$(k, k^\dagger) \in \left\{ (1, 0), (0, 1), (1, \infty), (\infty, 1), (\infty, \infty) \right\}.$$

Each of these possibilities can happen. The case $(k, k^\dagger) = (1, \infty)$ appears when ω is the free spanning forest of any graph G satisfying the conditions of Theorem 9.35, while $(\infty, 1)$ is the situation for the wired spanning forest. See [BLPS01] for more details. The other possibilities occur for Bernoulli percolation, as we shall see. Note that in light of Theorem 9.1 this requires us to rule out the cases $(0, 0), (1, 1), (0, \infty)$ and $(\infty, 0)$. The first two do occur in amenable planar graphs including $\mathbb{Z}^2$ in the critical percolation and for the uniform spanning tree respectively.

We separate the proof of Theorem 9.35 to several steps starting with the following one:

Theorem 9.36. *Let G be a transitive, nonamenable, planar graph with one end. Let ω be an invariant percolation on G. If ω has only finite components a.s., then $\omega^\dagger$ has infinite components a.s.*

The proof will use the following result from [BLPS99a], which we won't prove. It says that in certain graphs, if each edge is open with sufficiently high probability than there are infinite components (even if edges are not independent).

Theorem 9.37. *Let G be a unimodular nonamenable graph. There exists some $\epsilon >$ 0 such that if ω is an invariant percolation on G, and*

$$\mathbb{E}[\deg_\omega v] > \deg_G v - \epsilon,$$

then ω contains infinite clusters with positive probability.

Exercise 9.38. (Level 3) Show that this does not hold for $\mathbb{Z}^d$: Find an invariant percolation configuration with no infinite clusters where the expected degree is $2d - \epsilon$ for every $\epsilon > 0$.

Proof of Theorem 9.36. Suppose that both ω and $\omega^\dagger$ have only finite components with probability one. Thus given a component K of ω, there is a unique component K' of $\omega^\dagger$ which is closest to K that surrounds it almost surely. Similarly, for every component K' of $\omega^\dagger$, there is a unique component K'' of ω closest to K' that surrounds it almost surely. Let $\mathcal{K}_0$ denote the set of all components of ω, and inductively define

$$\mathcal{K}_{j+1} := \{K'' \ : \ K \in \mathcal{K}_j\}.$$

For $K \in \mathcal{K}_0$ let $r(K) := \sup\{j \ : \ K \in \mathcal{K}_j\}$ be the *rank* of K, and define $r(v) := r(K)$ where K is the component of v in ω. Note that $r(v)$ is a.s. finite. For each $s > 0$ let ω^s be the set of edges in $E(G)$ incident with vertices $v \in V(G)$ with $r(v) \leq s$. Then ω^s is an invariant bond percolation and

$$\lim_{s \to \infty} \mathbb{E}[\deg_{\omega^s} v] = \deg_G v.$$

Consequently, by Theorem 9.37, we find that ω^s has infinite components for all sufficiently large s with positive probability. This contradicts the assumption that ω and $\omega^\dagger$ have only finite components a.s. $\square$

The following Theorem which we won't prove here is from [BLPS99a] and [BLPS99b] in the transitive case. The extension to the quasi-transitive case is straightforward. Here we start with an invariant percolation configuration ω, and consider Bernoulli percolation on top of this random configuration.

Theorem 9.39. *Let G be a nonamenable quasi-transitive unimodular graph, and let ω be an invariant percolation on G which has a single component a.s. Then $p_c(\omega) < 1$ a.s.*

We will also need the following lemma from [BLPS99a].

Lemma 9.40. *Let G be a quasi-transitive nonamenable planar graph with one end, and let ω be an invariant percolation on G. Then the number of infinite components of ω is 0, 1 or ∞ with probability one.*

For the sake of completeness, we present a (somewhat different) proof of the last lemma here.

Proof. In order to reach a contradiction, assume that with positive probability ω has a finite number $k > 1$ of infinite components with positive probability and thus with probability one. Conditioning on this event, uniformly select a pair of distinct infinite components ω_1 and ω_2 of ω. Let ω_1^c be the subgraph of G spanned by the vertices outside of ω_1, and let τ be the set of edges of G that connect vertices in ω_1 to vertices in $\omega_1^c \cap \omega_2$. Set

$$\tau^\dagger := \{e^\dagger \ : \ e \in \tau\}.$$

From the definition we get that $\tau^\dagger$ is an invariant bond percolation in the dual graph $G^\dagger$. Using planarity, it is easy to verify that $\tau^\dagger$ has a.s. a bi-infinite path. This contradicts Theorem 9.39, and thereby completes the proof. □

Corollary 9.41. *Let G be a quasi-transitive nonamenable planar graph with one end, and let ω be an invariant percolation on G. Suppose that both ω and $\omega^\dagger$ have infinite components with probability one. Then a.s. at least one among ω and $\omega^\dagger$ has infinitely many infinite components.*

Proof. Draw G and $G^\dagger$ in the plane in such a way that every edge e intersects $e^\dagger$ in one point, v_e and there are no other intersections of G and $G^\dagger$. This defines a new graph $\hat{G}$, whose vertices are $V(G) \cup V(G^\dagger) \cup \{v_e \ : \ e \in E(G)\}$ and its edges are the edges of G and $G^\dagger$ "catted" in the appropriate points $\{v_e\}$. Note that $\hat{G}$ is also a quasi-transitive graph. Define a new percolation model on $\hat{G}$ by

$$\hat{\omega} := \left\{ \{v, v_e\} \in E(\hat{G}) : \begin{array}{l} v \in V(G), \ e \in \omega, \\ v \text{ is a vertex of } e \end{array} \right\} \bigcup \left\{ \{v^\dagger, v_e\} \in E(\hat{G}) : \begin{array}{l} v \in V(G^\dagger), \ e \notin \omega, \\ v \text{ is a vertex of } e^\dagger \end{array} \right\}$$

Informally, $\hat{\omega}$ corresponds to drawing simultaneously all edges of ω and $\omega^\dagger$. One can show that $\hat{\omega}$ is an invariant percolation on $\hat{G}$. Note that the number of infinite components of $\hat{\omega}$ is the number of infinite components of ω plus the number of infinite components of $\omega^\dagger$ and therefore contains at least two infinite components. Thus, by Lemma 9.40 applied to $\hat{\omega}$, we find that $\hat{\omega}$ has infinitely many infinite components which implies that either ω or $\omega^\dagger$ contains infinitely many infinite components. □

Proof of Theorem 9.35. By Lemma 9.40 both k and $k^\dagger$ are in $\{0, 1, \infty\}$ with probability one. The case $(k, k^\dagger) = (0, 0)$ is ruled out by Theorem 9.36. Since every two infinite components of ω must be separated by some component of $\omega^\dagger$, the situation $(k, k^\dagger) = (\infty, 0)$ is impossible. The same reasoning shows that $(k, k^\dagger) = (0, \infty)$ cannot happen. Finally the case $(k, k^\dagger) = (1, 1)$ is ruled out by Corollary 9.41. □

9.5.3 Bernoulli Percolation on Nonamenable Planar Graphs

We finish this section by applying and extending some of the previous results to the special case of Bernoulli percolation.

We start by extending the result proved in Theorem 9.35.

Theorem 9.42. *Let G be a quasi-transitive nonamenable planar graph with one end, and let ω be Bernoulli(p) bond percolation on G. Let k be the number of infinite components of ω, and k be the number of infinite components of $\omega^\dagger$. Then a.s.*

$$(k, k) \in \left\{ (1, 0), (0, 1), (\infty, \infty) \right\}.$$

Proof. By Theorem 9.35, it is enough to rule out the cases $(1, \infty)$ and $(\infty, 1)$. Let K be a finite connected subgraph of G. If K intersects two distinct infinite components of ω, then $\omega^\dagger \backslash \{e^\dagger : e \in E(K)\}$ has more than one infinite component. If $k > 1$ with positive probability, then there is some finite subgraph K such that K intersects two infinite components of ω with positive probability. Therefore, we find that $k^\dagger > 1$ with positive probability (since the distribution of $\omega^\dagger \backslash \{e^\dagger : e \in E(K)\}$ is absolutely continuous with respect to the distribution of $\omega^\dagger$). By ergodicity, this gives $k^\dagger > 1$ almost surely. An entirely dual argument shows that $k > 1$ with probability one whenever $k^\dagger > 1$ with positive probability.   □

Next we prove a connection between p_c of quasi-transitive nonamenable planer graphs with one end to p_u of its dual.

Theorem 9.43. *Let G be a quasi-transitive nonamenable planar graph with one end. Then $p_c(G) + p_u(G) = 1$ for Bernoulli bond percolation.*

Proof. Let ω_p be a Bernoulli(p) bond percolation on G. Then $\omega_p^\dagger$ is a Bernoulli($1 - p$) bond percolation on $G^\dagger$. It follows from Theorem 9.42 that the number of infinite components $k^\dagger$ of $\omega^\dagger$ is 1 when $p < p_c(G)$, ∞ when $p \in \big(p_c(G), p_u(G)\big)$ and 0 when $p > p_u(G)$. Thus $p_c(G^\dagger) + p_u(G) = 1 - p_u(G) + p_u(G) = 1$.   □

We can now finally prove the main theorems of this section:

Proof of Theorem 9.30. We start with the proof for bond percolation. Recall that if d is the maximal degree of the vertices in G then $p_c(G) \geq \frac{1}{d-1}$, see Theorem 4.6. By Theorem 9.32, ω_{p_c} has only finite components a.s. By Theorem 9.42, $\omega_{p_c}^\dagger$ has a unique infinite component a.s. Consequently, by Theorem 9.32 again, $\omega_{p_c}^\dagger$ is supercritical, that is, $p_c(G^\dagger) < 1 - p_c(G)$. An appeal to Theorem 9.43 now establishes the inequality $p_c(G) < p_u(G)$. Since $p_u(G) = 1 - p_c(G^\dagger) \leq 1 - \frac{1}{d^\dagger - 1}$, where $d^\dagger$ is the maximal degree of the vertices in $G^\dagger$, we get $p_u(G) < 1$, and the proof for bond percolation is complete. If ω is a site percolation on G, let ω^b be the set of edges of G with both endpoints in ω. Then ω^b is a bond percolation on G.

In this way, results for bond percolation can be adapted to site percolation. However, even if ω is Bernoulli, ω^b is not. Still, it is easy to check that the above proof applies also to ω^b.   □

Exercise 9.44. (Level 2) Verify that the proof above applies to site percolation by considering ω^b.

Proof of Theorem 9.31. By Theorem 9.43, $\omega_{p_u}^\dagger$ is critical Bernoulli bond percolation on $G^\dagger$. Hence, by Theorem 9.32, $\omega_{p_u}^\dagger$ has a.s. no infinite components. Therefore, it follows from Theorem 9.42 that ω_{p_u} has a single infinite component.   □

9.6   Product With $\mathbb{Z}$ and Uniqueness of Percolation

This subsection is based on a note with Gady Kozma (see [BK13]) where we showed that there exists a connected graph G with subexponential volume growth such that critical percolation on $G \times \mathbb{Z}$ has infinitely many infinite components.

Observe that if G is any connected graph and p is any number in $[0, 1]$, then the number of infinite clusters in p-percolation on $G \times \mathbb{Z}$ is deterministic, and is either 0, 1 or ∞. The proof is an easy consequence of the fact that one can take any finite set of vertices and translate it along the $\mathbb{Z}$ axis to get a disjoint set of vertices and thus independent edges.

In view of the last fact, Sznitman asked whether the Burton-Keane argument [BK89] applies. Namely, assume G is amenable, does it follow that $G \times \mathbb{Z}$ has only finitely many infinite clusters? As stated above the answer is negative. A binary tree with an infinite path added at the root serves as a counterexample.

We suggest a slight modification to the question:

Definition 9.45. A graph G is called strongly amenable if G doesn't contain nonamenable subgraph.

Exercise 9.46. (Level 3) Is $\mathbb{Z}^d$ strongly amenable?

Exercise 9.47. (Level 3) Determine whether your favorite graph is strongly amenable.

Open problem 9.48. *Assume G is strongly amenable, can one find an interval $[p_1, p_2]$ such that percolation on $G \times \mathbb{Z}$ has infinitely many infinite clusters for every p in this interval? What if we further assume that G has polynomial volume growth?*

We now describe without a proof an example of a strongly amenable graph of the form $G \times \mathbb{Z}$ with non uniqueness at p_c. We do not see yet any example of such a graph in which no percolation occurs at p_c. It is tempting to reformulate this question as $p_c = p_u$ but there is no monotonicity of uniqueness for graphs of the type $G \times \mathbb{Z}$. Indeed, connect the root of $\mathbb{Z}^{99}$ to the root of a 10 regular tree T and denote this graph by G. The parameters are chosen to satisfy

$$p_c(\mathbb{Z}^{100}) < p_c(T \times \mathbb{Z}) < p_u(T \times \mathbb{Z})$$

The first inequality follows from [Kes90] and the bound $p_c(T \times \mathbb{Z}) \geq \frac{1}{11}$ which holds for any graph whose degrees are bounded by 12. The second inequality follows from [Sch01]. It is not hard to see that on $G \times \mathbb{Z}$ for small p no percolation occurs. Then between $p_c(\mathbb{Z}^{100})$ and $p_c(T \times \mathbb{Z})$ there is a unique infinite cluster. Between $p_c(T \times \mathbb{Z})$ and $p_u(T \times \mathbb{Z})$ there are infinitely many infinite clusters. Finally, above $p_u(T \times \mathbb{Z})$ one has again a unique infinite cluster. This example can be generalized to an arbitrary, even infinite number of transitions.

Here is an example of a connected graph G with sub-exponential volume growth such that critical percolation on $G \times \mathbb{Z}$ has infinitely many infinite clusters.

Let d be some sufficiently large number fixed in the proof. The graph is constructed as follows. Take a tree of degree $4d$. Let $l_1 = 1$ and $l_{n+1} = l_n + \lceil d^2 \log(n+1) \rceil$. Now, for each $n \geq n_0$ (n_0 to be fixed in the proof) and for each edge (x, y) where x is in level $l_n - 1$ and y is in level l_n, disconnect (x, y) and instead take a copy of $\mathbb{Z}^d$ (considered as a graph with the usual structure) and connect x with the vertex $(0, \ldots, 0)$ and y with the vertex $(n, \ldots, n)$. All copies of $\mathbb{Z}^d$ (for all such (x, y)) are disjoint.

In the note with Gady, we show that at $p = p_c(\mathbb{Z}^{d+1})$ the graph $G \times \mathbb{Z}$ has infinitely many infinite clusters. One can rather easily convince oneself that in fact below p our graph $G \times \mathbb{Z}$ has no infinite clusters, so $p = p_c(G \times \mathbb{Z})$, but we will not do it here. Note that $p = \frac{1+o(1)}{2d}$ where $o(1)$ is as $d \to \infty$, see [Kes90] for more details.

Open problem 9.49. *Let G be an infinite graph with $p_c(G) = 1$ and look on Bernoulli percolation on $G \times \mathbb{Z}$. Does any infinite cluster intersects the fibers $\{v\} \times \mathbb{Z}$ infinitely often a.s.?*

Chapter 10
Percolation Perturbations

The unique infinite cluster is a perturbation of the underling graph that shares many of its properties (e.g. transience of random walk). Infinite clusters in the non uniqueness regime on the other hand admit some universal features which are not inherited from the underling graph, they have infinitely many ends and thus are very tree like. When performing the operation of contracting Bernoulli percolation clusters different geometric structures emerge. When the clusters are subcritical, we see random perturbation of the underling graphs but when the construction is based on critical percolation new type of spaces emerges. More precise definitions appears below. We end with an invariant percolation viewpoint on the incipient infinite cluster (IIC).

10.1 Isoperimetric Properties of Clusters

In this section we study expansion properties of infinite clusters of percolation on non-amenable graphs. The material in this section closely follows [BLS99].

Theorem 10.1. *Let G be a graph with a transitive unimodular closed automorphism group $\Gamma \subset Aut(G)$, and suppose that $h(G) > 0$. Let ω be a Γ-invariant percolation in G which is not empty (i.e., not empty with probability one)that has a.s. exactly one infinite component. Then (perhaps on a larger probability space) there is a percolation $\omega' \subset \omega$ such that $\omega' \neq \emptyset$ and $h(\omega') > 0$ a.s. Moreover, ω' can be chosen so that the distribution of the pair (ω', ω) is Γ-invariant.*

In the following, for $K \subset G$ a subgraph and $v \in V(G) \backslash V(K)$, we set $\deg_K v := 0$ and recall the following definitions from Chap. 8:

$$\alpha_K = \frac{1}{|V(K)|} \sum_{x \in V(K)} \deg_K(x),$$

I. Benjamini, *Coarse 3Geometry and Randomness*, Lecture Notes in Mathematics 2100,
DOI 10.1007/978-3-319-02576-6_10,
© Springer International Publishing Switzerland 2013

and

$$\alpha(G) = \sup\{\alpha_K \ : \ K \subset G \text{ is a finite subgraph}\}.$$

The proof of Theorem 10.1 is based on the following more quantitative result.

Theorem 10.2. *Let G be a graph with a transitive unimodular closed automorphism group $\Gamma \subset \text{Aut}(G)$ (in particular G is transitive). Assume ω is a Γ-invariant nonempty (i.e., not empty with probability one), percolation on G. If there exists $a > 0$ such that*

$$\mathbb{E}[\deg_\omega o \mid o \in \omega] > \alpha(G) + 2a, \tag{10.1}$$

then there is (perhaps on a larger probability space) a percolation $\omega' \subset \omega$ such that $\omega' \neq \emptyset$ and $h(\omega') \geq a$ with positive probability. Moreover, the distribution of the pair (ω', ω) can be chosen to be Γ-invariant.

Proof. Given any subgraph ω of G, we define a sequence of percolations ω_n on ω inductively as follows. First, let $\omega_0 := \omega$. Next assume that ω_n has been defined. Let β_n be a $\frac{1}{2}$-Bernoulli site percolation on G, independent of $\omega_0, \ldots, \omega_n$ and γ_n the union of the finite components K of $\beta_n \cap \omega_n$ which satisfy

$$\frac{|\partial_{E(\omega_n)} K|}{|K|} < a,$$

where $\partial_{E(\omega_n)} K$ denotes the set of edges of ω_n connecting K to its complement. We define $\omega_{n+1} := \omega_n \backslash \gamma_n$. Finally, define

$$\omega' := \bigcap_{n=0}^{\infty} \omega_n.$$

Note that ω' depends on ω, the parameter a and the independent Bernoulli percolations β_n.

Most of the proof will be devoted to showing that $\omega' \neq \emptyset$ with positive probability, but first we verify that $h(\omega') \geq a$. Indeed, let W be a finite nonempty set of vertices in G, F the set of all edges of G incident with W, and $F_0 \subset F$ a set such that $\frac{|F_0|}{|W|} < a$. To verify that $h(\omega') \geq a$ a.s., it is enough to show that the probability that $W \subset \omega'$ and $\omega' \cap F = F_0$ is zero. If $\omega_n \cap F = F_0$ for some n, then a.s. there is some $m > n$ such that W is a component of β_m. Now either $W \not\subset \omega_m$, in which case $W \not\subset \omega'$, or $W \subset \omega_m$, in which case W is not contained in ω_{m+1}, and thus not in ω'. On the other hand, if $\omega_n \cap F \neq F_0$ for every n, then also $\omega' \cap F \neq F_0$. Consequently $h(\omega') \geq a$ a.s.

Now set

$$D_n := \mathbb{E}[\deg_{\omega_n} o], \quad D_\infty := \mathbb{E}[\deg_{\omega'} o], \quad \theta_n := \mathbb{P}(o \in \omega_n), \quad \theta_\infty := \mathbb{P}(o \in \omega').$$

Our next goal is to prove the inequality

$$D_{n+1} \geq D_n - (\theta_n - \theta_{n+1})(\alpha(G) + 2a). \tag{10.2}$$

This will be achieved through use of the Mass-Transport Principle and the observation that

$$\theta_n - \theta_{n+1} = \mathbb{P}(o \in \gamma_n).$$

Fix n and define the random function $M : V(G) \times V(G) \to [0, \infty)$ as follows. For a vertex $v \in V(G)$, let $K(v)$ be the component of v in γ_n, which we take to be $\emptyset$ if $v \notin \gamma_n$. For $v, u \in V(G)$ we define

$$M(v, u) = \begin{cases} 0 & u \notin \gamma_n \\ \frac{\deg_{\omega_n} v}{|K(u)|} & u \in \gamma_n, v \in K(u) \\ \frac{1}{|K(u)|} \cdot |\{e \in \omega_n \ : \ e \text{ connects } v \text{ to } K(u)\}| & u \in \gamma_n, v \notin K(u) \end{cases}$$

Note that v and u need not be adjacent in order that $M(v, u) \neq 0$. Clearly, $\mathbb{E}[M(v, u)]$ is invariant under the diagonal action of Γ on $V(G) \times V(G)$. Consequently, the mass transport principle implies that

$$\sum_{v \in V(G)} \mathbb{E}[M(o, v)] = \sum_{v \in V(G)} \mathbb{E}[M(v, o)].$$

A straightforward calculation will show that

$$\sum_{v \in V(G)} M(o, v) = \deg_{\omega_n} o - \deg_{\omega_{n+1}} o ,$$

while, for $o \in \gamma_n$, we have $\sum_{v \in V(G)} M(v, o) = \frac{2}{|K(o)|} \cdot |\{e \in \omega_n \ : \ e \text{ is incident with } K(o)\}|$. The number of edges of G with both endpoints in $K(o)$ is at most $\alpha(G)|K(o)|/2$, and by construction, $|\partial_{\omega_n} K(o)| < a|K(o)|$. Hence

$$\sum_{v \in V(G)} M(v, o) < \alpha(G) + 2a \tag{10.3}$$

whenever $o \in \gamma_n$ and $\sum_{v \in V(G)} M(v, o) = 0$ otherwise. Therefore, $D_n - D_{n+1} = \mathbb{E}[\deg_{\omega_n} o - \deg_{\omega_{n+1}} o] = \sum_{v \in V(G)} \mathbb{E}[M(v, o)] \leq (\alpha(G) + 2a)\mathbb{P}(o \in \gamma_n) = (\alpha(G) + 2a) \cdot (\theta_n - \theta_{n+1})$, which is the same as (10.2).

An induction argument together with (10.2) gives

$$D_n \geq D_0 - \theta_0(\alpha(G) + 2a) + \theta_n(\alpha(G) + 2a).$$

Taking the limit $n \to \infty$ yields the inequality

$$D_\infty \geq D_0 - \theta_0(\alpha(G) + 2a) + \theta_\infty(\alpha(G) + 2a). \tag{10.4}$$

This gives $D_\infty > 0$, because (10.1) is equivalent to $D_0 - \theta_0(\alpha(G) + 2a) > 0$. Consequently, $\omega' \neq \emptyset$ with positive probability.   □

Remark 10.3. The following lower bound for θ_∞ is a consequence of (10.4) and the inequality $\theta_\infty \deg_G o \geq D_\infty$:

$$\mathbb{P}(o \in \omega') = \theta_\infty \geq \frac{D_0 - (\alpha(G) + 2a)\theta_0}{\deg_G o - (\alpha(G) + 2a)}$$

$$= \mathbb{P}(o \in \omega)\left(1 - \frac{\deg_G o - \mathbb{E}[\deg_\omega o \mid o \in \omega]}{h(G) - 2a}\right). \tag{10.5}$$

Proof of Theorem 10.1. Fix a base-point $o \in V(G)$. Let ω_* be the infinite component of ω. Conditioned on ω, for every vertex $v \in V(G)$, let $\phi(v)$ be chosen uniformly among the vertices of ω_* closest to v, with all $\phi(v)$ independent given ω. In addition for an edge $e = [v, u] \in E(G)$, let $\phi(e)$ be chosen uniformly among shortest paths in ω_* joining $\phi(v)$ to $\phi(u)$, with all $\phi(e)$ independent given $\{\phi(v)\}_{v \in V(G)}$ and ω. For an integer j, let η_j be the set of edges $e \in E(G)$ such that $\phi(e)$ is contained within a ball of radius j around one of the endpoints of e. Then $\eta_1 \subset \eta_2 \subset \cdots$ are Γ-invariant bond percolations on G with $\bigcup_j \eta_j = E(G)$. Consequently, $\lim_{j \to \infty} \mathbb{E}[\deg_{\eta_j} o] \to \deg_G o$. For each $k \in \mathbb{N}$, choose independently a random sample of ω defined as in the previous proof with $\{\eta_j\}_{j \geq k}$ replaced by γ_k and denote it by ξ_j. According to (10.5), we have that $\lim_{k \to \infty} \mathbb{P}(\xi_k \neq \emptyset) = 1$. Let $J := \inf\{j : \xi_j \neq \emptyset\}$. Then $J < \infty$ a.s. Set $\omega' := \phi(\xi_J)$. Since $h(\xi_J) \geq a$ a.s., we have also $h(\phi(\xi_J)) > 0$ a.s.   □

10.2 Contracting Clusters of Critical Percolation (CCCP)

Examine bond percolation on $\mathbb{Z}^d$. We define a new (multi)-graph by the following rule: contract each cluster of the bond percolation on $\mathbb{Z}^d$ into a single vertex and define a new edge between the clusters $\mathcal{C}, \mathcal{C}'$ for every closed edge that connected them in $\mathbb{Z}^d$. The result is a random graph G with high degrees (each vertex $v \in G$ belongs to a cluster $\mathcal{C}$ in $\mathbb{Z}^d$ and its degree is the number of closed edges coming out of $\mathcal{C}$). Of course, this can create double or multiple edges but this is not a problem (We can also think on G as a weighted graph instead of a multi-graph). When the percolation is *subcritical* one expects to see a perturbation of the lattice, but when the percolation is *critical* the random geometric structure obtained is expected to be rather different.

Here is a short summary of some results (with proof sketches) from a 2006 project with Ori-Gurel-Gurevich and Gady Kozma, which are not written yet, regarding the behavior at the critical value. We hope it will be enjoyable or at least useful to the readers.

Below we will examine G in $d = 2$ and in $d > 6$. We will often write *CCCP* instead of G (*CCCP* standing for Contracting Clusters of Critical Percolation).

10.2.1   Two Dimensional CCCP

Theorem 10.4. *CCCP has a single end with probability one.*

Proof. This follows immediately from the fact that for every vertex there exists clusters of arbitrary large size enveloping it.   □

Theorem 10.5. *CCCP is planar.*

Proof. Every quotient graph of a planar graph is planar. This follows from example from the fact the characterization of planarity using the non-existence of the minors K_5 and $K_{3,3}$.   □

Theorem 10.6. *CCCP has exponential volume growth.*

Proof Sketch. Examine a vertex $x \in \mathbb{Z}^2$. The boundary of a ball in CCCP is given by a path in the dual of the percolation on $\mathbb{Z}^2$. Because each scale contains such a path with probability $< 1 - c$, the *BK* inequality implies that you cannot have significantly more than 1 in each scale. In other words, the boundary of a ball of radius r in CCCP reaches to distance e^{cr} in $\mathbb{Z}^2$. This shows that the volume, measured in edges (and in this case it is the same in $\mathbb{Z}^2$ and in CCCP), grows exponentially. The volume measured in vertices also grows exponentially because CCCP has many vertices corresponding to a single vertex in $\mathbb{Z}^2$.   □

Similarly one may show that if $x, y \in \mathbb{Z}^2$ then the expected distance between their clusters in CCCP is $\approx \log |x - y|$.

Theorem 10.7. *CCCP is transient.*

Proof Sketch. One can construct a flow with finite energy by simply observing that the usual flow on $\mathbb{Z}^2$ (which has constant energy in each scale) and noting that only edges which have no cluster separating them from 0 contribute to the energy of the flow, and there are r^{2-c} of these at scale r with c as in the previous theorem.   □

Theorem 10.8. *The speed of a simple random walk on CCCP, measured in the Euclidean distance, is $\leq t^C$.*

Proof Sketch. Examine a cluster $\mathcal{C}$ surrounding the starting point at (Euclidean) scale r. This cluster has typically r^{β_1} edges coming out of it, where β_1 is the one-arm exponent. However, most of them go to "bubbles"-areas of CCCP which are only connected to the rest of CCCP through $\mathcal{C}$. Only r^{β_2} of them go to the parts of the graph connected to infinity, where β_2 is the two-arms exponent. Hence the walker spends a typical $r^{\beta_2-\beta_1}$ time just getting in and out of bubbles. It is known that $\beta_2 > \beta_1$ for any lattice, and in particular for $\mathbb{Z}^2$. This works also in continuous time because there are also r^{β_1} edges which go from the big cluster into a single-vertex cluster, from which the continuous time random talk takes time $\frac{1}{4}$ to escape.   □

We remark that $C \geq \frac{1}{2}$ because one can think about random walk on CCCP (say in continuous time) like random walk on $\mathbb{Z}^2$ which traverses open edges infinitely fast, and closed edges with the usual speed.

10.2.2   *d-Dimensional CCCP for d > 6*

Throughout this section we assume that critical percolation in our $\mathbb{Z}^d$ satisfies the triangle condition. By Hara and Slade [HS90] this holds if d is sufficiently large or if $d > 6$ and the lattice is sufficiently spread out.

Theorem 10.9. *CCCP has double-exponential volume growth.*

Proof Sketch. By Aizenman and Barsky [AB87] we have that $\mathbb{P}(|\mathcal{C}(x)| > n) \approx n^{-1/2}$. Hence when comparing $B(r)$ (the ball in CCCP) to $B(r+1)$ one can expect that $B(r+1) \approx |\partial B(r)|^2 \approx |B(r)|^2$. Justifying this formally requires to show that conditioning on $B(r)$ does not inhibit the growth of the clusters on its boundary, but this is quite standard in high-dimensional percolation.   □

Again, one may also show that if $x, y \in \mathbb{Z}^3$ then the expected distance between their clusters is $\approx \log\log|x - y|$.

Two lemmas which could be useful for the proof of such a result are the following:

Lemma 10.10. *For $x, y \in \mathbb{Z}^d$, the probability that they are connected in CCCP by $k \geq \frac{d}{2} - 2$ edges is $\approx |x - y|^{-4}$*

The probability seems small, but in fact this is the probability that both of them belong to large clusters, i.e. two clusters going to distance $\approx |x - y|$ (see [KN11]).

Proof Sketch. Denote by X the number of k-tuples of edges of $\mathbb{Z}^d$, $(e_1, \ldots, e_k)$ with $e_i = (x_i, y_i)$, such that $y_i \leftrightarrow x_{i+1}$ (connected) (defining y_0 and x_{k+1} our start and end points). Then a simple diagrammatic bound shows that $\mathbb{E}[X] \geq r^{2k+2-d}$ and $\mathbb{E}[X^2] \leq r^{4k+8-2d}$, if $k \geq \frac{1}{2}d - 2$. The diagram giving the main term for X^2 is (when $k < \frac{1}{2}d - 2$ a different diagram becomes the main term).   □

Lemma 10.11. *Let $a, b, c, d \in \mathbb{Z}^d$ such that the distance between every pair is $\approx r$. Let $k = \lceil \frac{d}{2} \rceil - 2$ and denote by $a \longleftrightarrow b$ the event that they are connected by k edges. Then*

$$\mathbb{P}(a \longleftrightarrow b, c \longleftrightarrow d) = \mathbb{P}(a \longleftrightarrow b)\mathbb{P}(c \longleftrightarrow d)(1 + O(r^{-1})).$$

Proof Sketch. The event that $a \longleftrightarrow b$, $c \longleftrightarrow d$ and the paths intersect can be estimated by diagrammatic bounds, and the main term comes from the diagram which has probability r^{4-2d}. Hence the lemma follows by the FKG and BK inequalities. □

Next we have the following estimation for the isoperimetric profile of CCCP:

Theorem 10.12. *The isoperimetric dimension of CCCP is d.*

Proof. Because it is a quotient graph of $\mathbb{Z}^d$ we immediately get that for any $A \subset$ CCCP, $|\partial A| \geq c|A|^{(d-1)/d}$ for an appropriate $c = c(d) > 0$ (here ∂A is the set of edges going out of A, and $|A|$ is the total number of edges between two vertices of A. With these definitions the inequality is an immediate consequence of the corresponding one for $\mathbb{Z}^d$). To show that one cannot get a better profile, take the ball $B(r) \subset \mathbb{Z}^d$ and remove from it any cluster that intersects its complement. The set one gets, F, is also a subset of CCCP because it is a collection of clusters. An edge in ∂F must have an open path from one of its endpoints to the outside of $B(r)$. Because the one-arm exponent is 2, the expected number of such edges in $B(s+1) \setminus B(s)$ is $s^{-2}r^{(d-1)/d}$ and summing over s we see that $\mathbb{E}[|\partial F|] \approx r^{(d-1)/d}$ while $\mathbb{E}[|F|] \approx r^d$. □

Theorem 10.13. *The speed of random walk on CCCP, measured in the Euclidean distance, is $\sqrt{t \log t}$.*

Proof Sketch. This should follow by an environment-as-viewed-from-the-particle argument but standard versions don't seem to apply directly. The idea is, though, that the particle spends approximately r^{-2} of the time in clusters with radius $\approx r$, so the contribution to the variance from them is ≈ 1. You get a contribution of 1 from every scale, and there are $\log t$ scales relevant at time t, so the variance should be $t \log t$. □

Theorem 10.14. *CCCP has no sublinear harmonic functions*

Proof Sketch. The walk has logarithmic entropy—this could follow from the previous theorem, but since it does not require the precise $\sqrt{t \log t}$, just a bound of t is enough. Therefore this follows from an environment viewed from the particle argument + Birkhoff's ergodic theorem. The theorem then follows from the results of [BDCKY11]. Applying this result literally will give this result in the chemical distance. A much stronger result could be stated in the Euclidean distance—here it would claim that any harmonic function h with $h(x) \leq C|x|/\log|x|$ where $|x|$ is the Euclidean distance, is constant. It should follow from the previous theorem (so this result is somewhat more dubious), and a slight rearrangement of [BDCKY11].   □

10.3   The Incipient Infinite Cluster (IIC)

The following is another model for a random rooted graph: We define the incipient infinite cluster (IIC) as a limit of critical percolation on a Cayley graph conditioned to survive to distance n, rerooted at a uniform vertex, where the limit is taken on a subsequential limit as in [BS01b].

More precisely we have the following. Start with a Cayley graph (one should mainly think about the case of $\mathbb{Z}^d$). For $n \in \mathbb{N}$ let G_n be the random cluster of on a-priori chosen vertex 0 conditioned to reach the boundary of the ball of radius n around it, together with a uniformly chosen vertex in the cluster which we call the root.

In the space of rooted graphs with uniformly bounded degree we can define a metric by $d((G,o),(G',o')) = 2^{-k}$ if and only if the ball of radius k around o in the graph G is isometric to the ball of radius k around o' in the graph G'. The metric d make this space into a compact space and therefore to any sequence we can find a converging subsequence.

Since the rooted graphs G_n belong to such a space we can take the limit of some subsequence G_{n_k} to which we call the IIC. For more details on this notion of convergence see the previous section.

Exercise 10.15. (Level 2) Show that in fact there is no need to pass to a subsequence, as the finite measures converges on any Cayley graph and in particular on $\mathbb{Z}^2$.

There are several equivalent definitions for IIC, for more information on the subject see [Jár03]. For example in the case $\mathbb{Z}^2$, one can define IIC by the following procedure: Perform critical percolation in a box, take the lowest spanning cluster, and pick uniformly a vertex on it as the root.

It is not hard to show that if there is no percolation at the critical parameter p_c then $p_c(\text{IIC}) = 1$ a.s. It is also likely that for any Cayley graph the incipient infinite cluster has a.s. at most a quadratic growth. As for now, we can't show that it has density zero even in $\mathbb{Z}^3$.

Having density zero for the IIC should be equivalent to $\theta(p_c) = 0$. Lets see this equivalence. If $\theta(p_c) > 0$, then clearly the IIC has a positive density. For the other direction we have the following:

Theorem 10.16. *If* $\theta(p_c) = 0$ *then the IIC has density* 0.

Proof. Suppose $\theta(p_c) = 0$ and fix some $\epsilon > 0$. Consider the event $A(L,\epsilon)$ that in the ball $B(L)$ of side length L centered at 0 a proportion of at least ϵ of the vertices is connected to the boundary of $B(L)$. Then $P(A(L,\epsilon))$ tends to zero as

L tends to infinity. Choose L large enough so that $P(A(L, \epsilon)) < \frac{\epsilon}{2}$. Now let k be a large positive integer and consider the box $B(kL)$ tiled by translates of $B(L)$ in the obvious way. In order for at least a proportion of 2ϵ of the vertices in $B(kL)$ to be connected to the boundary of $B(kL)$ it is necessary that at least a proportion of ϵ of the boxes of side length L tiling $B(kL)$ have at least a proportion of ϵ of their vertices connected to their respective boundaries. These are independent events, each having a probability of at most $\frac{\epsilon}{2}$. Therefore, the probability that more than an ϵ-proportion of them occurs decays exponentially in the number of trials. Consequently, $P(A(kL, 2\epsilon)) < \exp(-ck^d)$ for some $c > 0$. Now, if we condition on the outside of $B(kL)$ under the IIC, the conditioned probability that 0 connects to the boundary of $B(kL)$ is at least $\exp(-Ck)$ for some constant $C = C(L)$. We can therefore choose k so large that ck^d is much larger than $C \cdot k$. In that case, the conditioned probability of $A(kL, 2\epsilon)$ under the IIC is arbitrarily small, which proves the claim.   □

We therefore know that $p_c(IIC) = 1$, and that it has density 0. We also believe, at least in the case of Cayley graphs, that the IIC has at most internal quadratic growth (i.e. quadratic growth in the induced metric of the IIC graph. Perhaps by assuming nonamenability one can show that the IIC is small? Since the IIC is an unbiased limit in the sense of [BS01b], it is a unimodular random graph. This is interesting, because of course it does not have translation invariance of any form. But unimodularity can replace that in many arguments, see [AL07]. How does IIC looks like on a $2 - 3$-tree? Does it have linear growth there? Assuming $\theta(p_c) = 0$ then by the BK inequality, there are no two disjoint paths to infinity a.s. The fact that there are no two disjoint paths to infinity already implies that there are infinitely many cut-points. The internal growth seems harder to deal with than the external growth (growth with respect to the distances in the original graph). For example, the external growth rate of the IIC on the triangular lattice is $r^{2-5/48+o(1)}$ but the internal growth is still unknown.

Theorem 10.17. *On a nonamenable Cayley graph the probability that a random walk started at the origin meets the IIC at some time $t > 0$ is not one.*

This is a form of small density result and is based on an unpublished idea of Oded Schramm.

Proof. Suppose that the random walk meets the IIC again with probability 1. Sample a random walk $S(t)$ up to its first visit with the IIC at some time $t > 0$. Then translate the IIC so that $S(t)$ is translated to the root. Unimodularity of the IIC implies that the final configuration is again the IIC. This implies that the random walk meets the IIC infinitely often (assuming it revisits the IIC with probability 1). Now pick a large k and consider the branching random walk that branches into two every k-th time it meets the IIC and branches into 3 at time $t = 0$. The value of k needs to be sufficiently large so that this BRW is transient. This defines a mapping of the 3-regular tree into the IIC. Now unimodularity shows that the pull back to the tree of the internal distance function on the IIC has an invariant law. Thus one can define an invariant percolation on the tree by removing all edges corresponding to internal distance larger than some fixed constant, and if the constant is large enough the expected degree of the root is larger than 2. By Olle Häggström 's theorem [Häg97], the invariant percolation has an infinite component and in fact $p_c < 1$ with positive probability. This implies that $p_c(IIC) < 1$. However, since a.s. there are infinitely many cut-points, it follows that $p_c(IIC) = 1$ a.s., which gives a contradiction.   □

10.3.1   *Discussion and Questions*

One can ask if the IIC is intrinsically tail trivial (its embedding in the ambient graph is not tail trivial in a hyperbolic Cayley graph). One can also ask if it is ergodic, in the sense that the basepoint invariant events all have probability 0 or 1. It should be ergodic. Even without proving ergodicity, we can pass to an ergodic component and since an ergodic component is a.s. unimodular, we conclude that the probability the RW revisits the IIC under the ergodic component measure is less than 1. Stationarity of the IIC viewed from the times in which the walk visits it then implies that the probability that the SRW meets the IIC infinitely often is 0.

It would also be nice to show that $P(v \in IIC)$ goes to zero as $d(v, o)$ goes to infinity (which is a good form of zero density for the IIC). However, at the moment the above only gives $P(S(t) \in IIC)$ goes to zero as t goes to infinity where $S(t)$ is as before a sample of a simple random walk. Does these cut-point drift linearly to infinity in the external distance, in non-amenable Cayley graphs? It follows that for any unimodular random subgraph of a non amenable Cayley graph with $p_c = 1$, random walk will not hit it infinitely often. Are there such graphs with internal exponential volume growth? Yes, there are. If you take the rooted 3-regular tree embedded randomly in a 4-regular tree, the SRW will meet it only finitely many times a.s. Note that infinitely many cut edges also implies recurrence.

A line can have exponential external growth, but is there a unimodular random graph with infinitely many cut points with exponential internal growth? The 3-regular tree has an exponential growth unimodular subgraph with infinitely many cut points: Take the 3-regular tree, choose an end, and consider the levels of the tree with respect to the given end. Take the tree spanned by the set of vertices above a certain level (above refers to the end being at the top), say level zero, and choose the root to be in level k with probability 2^{-k}. This is a unimodular rooted graph (also the limit of n-level binary trees with root chosen uniformly). You can embed it in the rooted d-regular tree ($d \geq 3$) randomly uniformly with root going to root.

Could something be proved about the distance between (internal or external) cutpoints in the IIC? If so, can one say something about the distance to 0 of the nth cut point? Again, at least in nonamenable groups, can we show that the IIC has linear speed? Is it possible to show it for hyperbolic planar groups via the SRW theorem above?

At least the fact that SRW will not hit it infinitely often mean that in the planar hyperbolic Cayley case the IIC converge radially. The planar hyperbolic setting is definitely a familiar setting in which there is more to work with. Jarai [Jár03] showed that in $\mathbb{Z}^2$ there is a coupling of the IIC to be contained in supercritical Bernoulli percolation, is this so in $\mathbb{Z}^3$, and on any Cayley graph or even on any graph?

Chapter 11
Percolation on Expanders

This section is devoted to percolation on finite graphs. More precisely we will try to understand percolation on a sequence of finite graphs, whose number of vertices tends to infinity. Detailed proofs of the material appearing in this section and additional extensions can be found at [ABS04].

Definition 11.1. Let $G_n = (V_n, E_n)$ be a sequence of d-regular connected finite transitive graphs such that $V_n \to \infty$. Recall that the finite graph formulation of Cheeger constant is given by

$$h(G) = \inf_{\substack{0 \le |S| \le |V_n|/2 \\ S \subseteq V_n}} \frac{|\partial S|}{|S|}.$$

We say that a the sequence G_n is an expander if there exists some $c > 0$ such that $h(G_n) > c$ for every $n \in \mathbb{N}$. In several cases we also allow a sequence of non regular graphs to be expander provided that the degrees of their vertices are uniformly bounded.

For an example of an explicit expander construction and much more on the subject, see [HLW06]. A random matching in a cycle, see Exercise 1.10, is an expander with high probability, though not a regular one.

We turn to introduce two metrics on graphs that are particularly relevant in the case of expanders.

Definition 11.2. Let $G = (V, E)$ be a finite graph.

- The *diameter* of G is the quantity $\sup_{u,v \in V} d_G(u, v)$.
- The *girth* of the graph, denoted girth(G), is the minimum length of a cycle in G.

In the following exercises, expanders are characterized by small diameters and rapid covering by small neighborhoods.

I. Benjamini, *Coarse 3Geometry and Randomness*, Lecture Notes in Mathematics 2100, 97
DOI 10.1007/978-3-319-02576-6_11,
© Springer International Publishing Switzerland 2013

Exercise 11.3. (Level 1) Let $G = (V, E)$ be a d-regular graph with diameter D. Show that $D \leq c \log |V|$, where c is a constant that depends on $h(G)$ and d.

Exercise 11.4. (Level 2) Let $G = (V, E)$ be a graph. Show that if $S \subseteq V$, satisfies $|S| \leq \frac{|V|}{2}$, then $|V \setminus S^k|$ decays exponentially with k, where S^k is the kth neighborhood of S, i.e., $S^k = \{u \in V, d(u, S) \leq k\}$.

Exercise 11.5. (Level 1) Show that $\text{girth}(G) \leq 2 \log_{d-1} |V|$ in every d-regular graph $G = (V, E)$.

Exercise 11.6. (Level 3) Consider the cycle with random matching for d-regular graphs. Show that around a randomly chosen vertex the graph looks like a tree up to depth $c \cdot \log n$.

Open problem 11.7. *Let G be an expander. Assume further that there is a vertex $v \in G$ such that percolation $p = \frac{1}{2}$ on G satisfy*

$$P_{\frac{1}{2}} \left(\text{the connected component of } v \text{ has diameter} > \frac{\text{diam}(G)}{2} \right) > \frac{1}{2}.$$

Show that there $p = \frac{1}{2}$ percolation on G contains a giant component with high probability.

The Erdös-Rényi Model

The Erdös-Rényi model is a random graph model we now describe.

Definition 11.8. Given a function $p : \mathbb{N} \to [0, 1]$, whose value at n we denote by p_n, define a random graph $G(n, p_n) = (V, E)$ as follows: V is a set of n vertices and for each $u, v \in V$, the edge $\{u, v\}$ is declared to be an edge in E independently with probability p_n.

The Erdös-Rényi model exhibits some phenomenons which mimic those of percolation in finite graphs. More precisely we have the following phase transition:

Theorem 11.9. *Assume $n \cdot p_n$ tends to a constant $c > 1$, then with probability converging to 1 as $n \to \infty$, $G(n, p_n)$ will have a unique giant component, i.e., a component containing a positive fraction of the vertices. In addition, no other component will contain more than $O(\log n)$ vertices. On the other hand if there exists a constant $0 < c < 1$ such that $np_n < c$, then with probability converging to 1 as $n \to \infty$, $G(n, p_n)$ doesn't contain a giant component.*

In the next two sections, we will study a generalization of this model to expanders and discuss two main theorems from [ABS04]:

• Existence of a giant component in a d-regular expander family with girth tending to infinity.
• Uniqueness of the giant component in an expander family with a uniformly bounded degree.

The giant component with high probability has the structure of an expander decorated by small graphs with exponential tail on their size.

11.1 Existence of a Giant Component

The goal of this subsection is to study the first point presented above. More formally we will prove the following theorem:

Theorem 11.10. *Let G_n be an expander family of finite d-regular graphs such that $h(G_n) > c > 0$ for every $n \in \mathbb{N}$. If girth$(G_n) \to \infty$ as $n \to \infty$, then:*

1. For every $p > \frac{1}{d-1}$ there exists $\alpha > 0$ such that

$$\mathbb{P}_p \text{ (there is an open cluster of size } \geq \alpha|G_n|) \xrightarrow[n \to \infty]{} 1.$$

2. If $p < \frac{1}{d-1}$ then $\forall \alpha > 0$

$$\mathbb{P}_p \text{ (there is an open cluster of size } \geq \alpha|G_n|) \xrightarrow[n \to \infty]{} 0.$$

We first introduce a very useful and commonly used technique developed by Erdös and Rényi. The name of the technique is "sprinkling" or "staged exposure" and we will use it to prove the last theorem. In fact we already used sprinkling in the previous section.

Consider percolation on a graph G where each edge is open with probability p (independently of the other edges). Now consider the case where we perform two rounds of exposure. Let p_1 and p_2 be two positive reals such that $1 - p = (1 - p_1)(1 - p_2)$. The "staged exposure" consists of two steps: the first is percolation with probability p_1 and the second is percolation with probability p_2 on the edges that remained closed after the first step (in each round of exposure the edges are picked to be open or closed independently of each other). It is easy to see that the final result of the two rounds of exposure is equivalent to the result of percolation with edge probability p.

We now outline the proof of Theorem 11.10.

Proof Sketch of Theorem 11.10. We start the proof by dealing with the second part which is easier. In fact the proof of the second part does not use the assumptions that G_n is an expander family and that girth$(G_n) \to \infty$ as $n \to \infty$. By Theorem 4.6 if G is an infinite d-regular graph, then $p_c(G) \geq \frac{1}{d-1}$. Notice that this statement is exactly the infinite version of part 2 of our theorem, and the proof is very similar.

If there is an open cluster of size $\geq \alpha|G_n|$ for some $\alpha > 0$ then there must be an open self avoiding path of length $\beta \log(|G_n|)$ for some $\beta > 0$ (this follows from the fact that the degree is uniformly bounded by d). However, for $p < \frac{1}{d-1}$ the probability that there exists such a path tends to 0 as n tends to ∞. For a more detailed version see the proof of Theorem 4.6.

Next we outline the proof of the first part of the theorem. Let $p = \frac{1+\epsilon}{d-1}$ and $p_1 = \frac{1+\frac{\epsilon}{2}}{d-1}$. We define p_2 to be the unique solution to the equation $1-p = (1-p_1)(1-p_2)$ (this implies $p_2 > \frac{\epsilon}{2d}$). For every vertex $v \in G_n$ the $\frac{\text{girth}(|G_n|)}{2}$-neighborhood of v looks exactly like a d-regular tree. Let us perform the first stage of exposure with probability p_1. Because $p_1 > \frac{1}{d-1}$, by standard branching process argument, with high probability there exists $b \in (0, 1]$ such that $bn \pm o(n)$ of the vertices lies in clusters, each of them is of size at least $\frac{\text{girth}(|G_n|)\epsilon}{6}$. If $f(n) \geq \gamma n$ for some $\gamma > 0$ then each cluster of size $\frac{\text{girth}(|G_n|)\epsilon}{6}$ is also an open cluster of linear size. Thus it is enough to deal with the case $\text{girth}(|G_n|) \ll n$. We shall refer to open clusters of size at least $\frac{\text{girth}(|G_n|)\epsilon}{6}$ as big clusters. The number of such big clusters is at most $\frac{bn \pm o(n)}{\text{girth}(|G_n|)\epsilon/6}$. Next perform the second stage of exposure with parameter p_2. We will show that a lot of the big components merge into a component of size $\gamma|G_n|$ for some $\gamma > 0$. To show this we will use the fact that we are dealing with an expander family of graphs. The proof will be completed once we prove the following claim, since it implies that there must be an open cluster of size at least $\frac{2bn}{3}$. In order to apply the claim we split the big clusters into two parts, satisfying the conditions of the claim, with no path between them in contradiction to the claim   □

Claim 11.11. *There is no way to split the big components into two parts A and B, each containing at least $\frac{bn}{3}$ vertices, in a way that there are no open paths between A and B.*

Proof. Using the fact that $h(G_n) > c > 0$, we know that any cut set between A and B is of size at least $\frac{cbn}{3}$. Menger's theorem implies that there are at least $\frac{cbn}{3}$ pairwise edge-disjoint paths between A and B in G_n. Since G_n has only $\frac{dn}{2}$ edges, at least half of these pairwise edge-disjoint paths are of length at most $\frac{3d}{cb}$. We shall refer to these paths as short. Therefore there are at least $\frac{cbn}{6}$ short paths. We now prove that there must be an open short path between A and B.

$$\mathbb{P}\,(\nexists \text{ an open path between } A \text{ and } B)$$

$$= \mathbb{P}\,(\text{all paths between } A \text{ and } B \text{ are closed})$$

$$\leq \mathbb{P}\,(\text{all short paths between } A \text{ and } B \text{ are closed})$$

$$= \mathbb{P}\,(\text{a certain short path is closed})^{\text{the number of short paths}}$$

$$= (1 - \mathbb{P}\,(\text{a certain short path is open}))^{\text{the number of short paths}} \qquad (11.1)$$

$$\leq (1 - p_2^{\frac{3d}{cb}})^{\frac{cbn}{6}} \leq \exp\left(-p_2^{\frac{3d}{cb} \cdot \frac{cbn}{6}}\right)$$

$$\leq \exp\left(\left(-\frac{\epsilon}{2d}\right)^{\frac{dn}{2}}\right) = o(2^{-(\frac{bn \pm o(n)}{\epsilon f(n)/6})})$$

As there are at most $\frac{bn \pm o(n)}{\varepsilon f(n)/6}$ big open clusters, there are at most $2^{\left(\frac{bn \pm o(n)}{\varepsilon f(n)/6}\right)}$ ways to choose such A and B (i.e. to partite the big open clusters into two parts). By the union bound we obtain that the claim holds for all such A and B, thus proving the claim.   □

11.2 Uniqueness of the Giant Component

Next we turn to consider the uniqueness of the giant component. We will show that there is at most one giant component when percolating on an expander family of graphs.

Theorem 11.12. *Let $G_n = (V_n, E_n)$ be an expander family with uniform maximal degree d. Fix some $\varepsilon > 0$ and assume that $\varepsilon \le p_n \le 1 - \varepsilon$. Then for every $c > 0$*

$$\mathbb{P}\left(G_n(p_n) \text{ contains more than one component of order at least } c|V_n|\right) \xrightarrow[n \to \infty]{} 0.$$

The proof uses the notion of up-sets which are quite similar to monotone boolean functions.

Definition 11.13. Let E be a finite set. A subset $\mathcal{X}$ of $\mathcal{P}(E)$ (the family of subsets of E) is called an up-set, if $A \in \mathcal{X}$ and $A \subset B \subseteq E$, implies $B \in \mathcal{X}$.

Definition 11.14. Let f be a boolean function (i.e. $f : \{0, 1\}^m \to \{0, 1\}$).

- f is called balanced if

$$|\{x : f(x) = 1\}| = |\{x : f(x) = 0\}| = 2^{m-1}$$

- For $x, y \in \{0, 1\}^m$ we say that $x <' y$ if $x_i = 1$ implies $y_i = 1$ for every $1 \le i \le m$.
- f is called monotone if $f(x) = 1$ and $x <' y$ implies $f(y) = 1$.

Remark 11.15. In fact, we can define monotone functions in a more general setting. Let $(A, <_A)$, $(B, <_B)$ be two ordered sets. A function $f : A \to B$ is called monotone if $x, y \in A$ and $x <_A y$ implies $f(x) <_B f(y)$.

One natural way to define up-sets is via monotone functions. Given a set E we can define an order $<_E$ on $\mathcal{P}(E)$ by inclusion. If $f : \mathcal{P}(E) \to \mathbb{N}$ is a monotone function, where on $\mathbb{N}$ we take the standard ordering, then for every $i \in \mathbb{N}$ the family $\{X \in \mathcal{P}(E) : f(X) \ge i\}$ is an up-set.

Definition 11.16. Let E be a finite set, and let $\mathcal{A} \subseteq \mathcal{P}(E)$ be an upset. Given $A \in \mathcal{P}(E)$, we say that $e \in E$ is A-pivotal (w.r.t $\mathcal{A}$) if $A^e := A \cup \{e\} \in \mathcal{A}$ but $A_e := A \setminus \{e\} \notin \mathcal{A}$.

Theorem 11.17. *Let E be a finite set, and $\mathcal{A} \subseteq \mathcal{P}(E)$ an upset. Let $A \subseteq E$ be obtained by selecting each element of E independently with probability p, where $p \in [x, 1 - x]$ (for some $x \in (0, 1/2)$), and let $e \in E$ be chosen uniformly at random and independently of the choice of A. Then there exists some $\alpha > 0$ such that*

$$\mathbb{P}(e \text{ is } A\text{-pivotal w.r.t } \mathcal{A}) \leq \frac{\alpha}{\sqrt{|E|}}$$

Proof. Given E and $p \in [x, 1 - x]$ we will construct a pair (A, e) as follows. Let $e_1 < e_2 < \ldots < e_k$ be a random ordering of E chosen uniformly. Let X be a random variable that is distributed according to binomial distribution with parameters k and p, i.e., $X \sim B(k, p)$. We take A to be the first X elements of E, $A = \{e_1, e_2, \ldots e_X\}$. In addition, we choose e to be $e = e_X$ with probability $\frac{X}{k}$ or $e = e_{X+1}$ with probability $\frac{k-X}{k}$. We claim that this pair is distributed according to the theorem's conditions for A and e. The fact that the marginal distribution of A is correct is immediate. The edge e needs to be chosen uniformly and independently from A. Given a set of elements A, it could have been arisen from $|A|!(k - |A|)!$ orderings, each with equal probability. Therefore each element of A has the same probability to be the last element, which is $\frac{1}{|A|}$. Thus, the probability that any element of A will be the random element e is $\frac{1}{|A|} \frac{|A|}{k} = \frac{1}{k}$. Similarly, again conditional on the choice of A, any element outside A also has a chance of $\frac{1}{k}$ to be equal to e.

Given an ordering $e_1, \ldots, e_k$, let $A_l = \{e_1, \ldots, e_l\}$ for $1 \leq l \leq k$. Since A is an up-set there will be exactly one l such that $A_l \notin \mathcal{A}$ but $A_{l+1} \in \mathcal{A}$. Thus, e is A-pivotal if $X = l$ and $e = e_{l+1}$ or if $X = l + 1$ and again $e = e_{l+1}$. This happens with probability (conditional on the ordering)

$$\frac{k - l}{k} \mathbb{P}_p(X = l) + \frac{l + 1}{k} \mathbb{P}_p(X = l + 1) \leq \frac{l + 1}{k} \max_{p, m} \mathbb{P}_p(X = m)$$

The maximum is taken over $p \in [x, 1 - x]$ and $0 \leq m \leq k$. This bound is independent of the ordering and therefore it is also a bound for the probability that e is A-pivotal. It is not hard to show that this is bounded from above by some constant over $\sqrt{k}$. which complete the proof. $\qquad\square$

We now turn to prove Theorem 11.12, but first we need one more definition.

Definition 11.18. Given a subgraph H of G_n and $c > 0$, we say that an edge $e \in G_n$ is an L-*bridge* if H_e (H without e) contains two large components, each one with more than $c|V_n|$ vertices, which are connected by e.

Proof of Theorem 11.12. Given a set of edges F in G_n, we define

$$Z(F) = \frac{Y(F)}{c|V_n|} - L(F),$$

where $L(F)$ is the number of large components in the graph generated by F, and $Y(F)$ is the number of vertices in the large components of the graph generated by F. Note that $Z(F)$ is increasing with respect to inclusion, since adding a single edge to F, may increase $L(F)$ by at most one, and if $L(F)$ indeed changed by adding an edge, then $Y(F)$ must have increased by at least $c|V_n|$. Therefore, for any t we have that $\mathcal{A}_t = \{F \in E(G_n) : Z(F) \geq t\}$ is an up-set, and an L-bridge must be pivotal for at least one of the $\mathcal{A}_t$'s. Using Theorem 11.17 we have for an appropriate $\alpha > 0$

$$\mathbb{P} \left(e \text{ is an } L \text{ bridge} \right) \leq \left(\left\lfloor \frac{1}{c} \right\rfloor - 1 \right) \frac{\alpha}{\sqrt{|E_n|}},$$

when e is chosen uniformly. Note that we used the fact that the maximal value of $Z(F)$ is $\left(\left\lfloor \frac{1}{c} \right\rfloor - 1 \right)$. Since c is a constant and G_n has a uniformly bounded degree, it is possible to reformulate this as

$$\mathbb{P} \left(e \text{ is an } L \text{-bridge} \right) \leq \frac{\beta}{\sqrt{|V_n|}}.$$

Since G_n is an expander, we have that for any large enough set A in V_n, there is some r such that the vertices at distance at most r from A have cardinality at least $\frac{3}{4}|V_n|$. This means that for two large sets, there must be many paths of a constant length connecting them. Recall that the degree of G_n is uniformly bounded by d. Thus, there are at most d^r edges at distance at most r from any specific vertex. Choosing v uniformly gives

$$\mathbb{P} \left(\begin{array}{c} \text{There is an } L\text{-bridge at} \\ \text{distance at most } r \text{ from } v \end{array} \right) \leq \sum_e \mathbb{P} \left(\begin{array}{c} e \text{ is at distance at} \\ \text{most } r \text{ from } v \end{array} \right) \mathbb{P} \left(e \text{ is an } L\text{-bridge} \right)$$

$$\leq \max_e \mathbb{P} \left(\begin{array}{c} e \text{ is at distance at} \\ \text{most } r \text{ from } v \end{array} \right) \sum_e \mathbb{P} \left(e \text{ is an } L\text{-bridge} \right)$$

$$\leq \frac{d^r}{|V_n|} |E_n| \frac{\beta}{\sqrt{|V_n|}}.$$

Since we also know that $|E_n| \leq d|V_n|$, it follows that there exists some constant $\gamma > 0$ (for r constant, as defined previously) such that

$$\mathbb{P} \left(\begin{array}{c} \text{There is an } L\text{-bridge at} \\ \text{distance at most } r \text{ from } v \end{array} \right) \leq \frac{\gamma}{\sqrt{|V_n|}}.$$

If there are two vertices belonging to two different large components, both of them in distance at most r from some $v \in V_n$, by adding at most $2r$ edges we obtain an L-bridge at distance at most r from v. Selecting v randomly and changing the percolation on the edges along the path gives

$$\mathbb{P}\left(\begin{array}{c}\text{There are two large components}\\\text{at distance at most } r \text{ from } v\end{array}\right) \le (pd)^{-2r}\mathbb{P}\left(\begin{array}{c}\text{There is an } L\text{-bridge at}\\\text{distance at most } r \text{ from } v\end{array}\right),$$

where p is the probability for an edge to be open. Since r was chosen so that for each of the large components, $\frac{3}{4}$ of the vertices lie at distance at most r from it, for every pair of large components, at least $\frac{1}{2}$ of all vertices must lie at distance at most r from both of them. Thus, choosing a vertex v randomly should give two vertices at distance r from v belonging to two distinct components with probability at least $\frac{1}{2}$, so there are two large components with probability at most

$$\frac{2\gamma}{(pd)^{2r}\sqrt{|V_n|}}$$

which goes to zero as n goes to infinity. $\square$

Remark 11.19. Beyond the expanders families and the family of the lattice tori not much is known regarding percolation on finite graphs. Rather simple conjectures are still open, [ABS04] contain several questions and conjectures regarding percolation on other finite graphs, in particular vertex transitive ones.

11.3 Long Range Percolation

The long range percolation (LRP) is a random graph defined by three parameters $n \in \mathbb{N}$, $s > 0$ and $\beta > 0$ defined as follows: The vertex set is $V = \{0, \ldots, n\}$ and the random edge set is defined by letting each pair $\{i, j\}$ where $0 \le i, j \le n$ independently with probability $\beta|i - j|^{-s}$. Let D be the diameter of the random graph obtained in this model. Observe that D decreases as a function of s (the graph becomes sparser) and that as s gets smaller, the random graph becomes closer to an expander. See [Bis04] for additional background and references.

The following graph summarize the behavior of D as a function of n for different values of s

s	Rough behavior of D
$s > 2$	Constant, with value that is not known times n
$s = 2$	$n^{\theta(\beta)}$, where $0 < \theta(\beta) < 1$.
$1 < s < 2$	Polylogarithmic in n
$s = 1$	$\frac{\log(n)}{\log\log(n)}$
$s < 1$	Uniformly bounded.

The case $s = 2$ is relatively new and can be found in [DS13].

Exercise 11.20. (Level 3) Show that when $s > 2$, with high probability the LRP graph has linear diameter.

Open problem 11.21. *It was not shown yet that $\theta(\beta)$ is continuous or if it is monotone and there is no even a guess for it. A non trivial scaling limit should exist when $s = 2$.*

To our knowledge, beyond simple random walk, no other random processes (e.g. particle systems) were studied on long range percolation clusters.

Chapter 12
Harmonic Functions on Graphs

The main goal of this section is to present the Kaimanovich-Vershik entropic criterion for the existence of harmonic function on Cayley graphs. Note that this section requires more background in probability compared to previous sections. We begin with some definition and simple facts.

12.1 Definition

Definition 12.1. Let $G = (V, E)$ be a locally finite graph. A function $h : V \to \mathbb{R}$ is called harmonic if for every $v \in V$,

$$h(v) = \frac{1}{\deg(v)} \sum_{(u,v) \in E} h(u).$$

The following theorem is a corollary to the entropy approach below.

Theorem 12.2. *If G is a vertex transitive non amenable graph then G admits non constant bounded harmonic function.*

Dirichlet Problem. Problem: Given a finite connected cluster C in a graph G with values on the boundary, construct an harmonic function on the cluster which coincide with the values on the boundary.

Solution: Fix v inside the cluster and start a simple random walk $\{X_n\}_{n \geq 0}$ on the graph G from v. We denote X_{T_v} the position at which the simple random walk that starts from v hits the boundary of the cluster C. Now define

$$h(v) = \mathbb{E}\left[f\left(X_{T_v}\right)\right].$$

Then h solves Dirichlet problem. In fact we have the following:

I. Benjamini, *Coarse 3Geometry and Randomness*, Lecture Notes in Mathematics 2100, DOI 10.1007/978-3-319-02576-6_12, © Springer International Publishing Switzerland 2013

Proposition 12.3. *The function h is the unique solution of the Dirichlet problem.*

Proof Sketch. For the existence, we just have to check that the function h indeed defines an harmonic function inside C: This follows from the Markov property of the simple random walk applied at time 1. The uniqueness follows from the maximum principle. The maximum principle says that if an harmonic function has a maximal value then this value is archived on the boundary. Assume that there exist two solutions of the Dirichlet problem. Their difference, which we denote by h, is another harmonic function which vanish on the boundary—hence by the maximum principle applied to h and $-h$ this function must be the constant zero function.

The following exercises give some examples for graphs with and without non constant harmonic functions:

Exercise 12.4. (Level 1) Show that the space of bounded harmonic function on the 3-regular tree is infinite dimensional.

Exercise 12.5. (Level 1) Show that on $\mathbb{Z}$ there are no bounded harmonic function but the constants.

Exercise 12.6. (Level 2) Show the same for $\mathbb{Z}^2$.

Exercise 12.7. (Level 3) Show the same for $\mathbb{Z}^d$ for every $d \geq 3$ (little harder, use coupling)

In fact one can prove much more general fact about harmonic function on $\mathbb{Z}^d$:

Example 12.8. There is no sublinear harmonic function on $\mathbb{Z}^d$, that is if $f : \mathbb{Z}^d \to \mathbb{R}$ is harmonic such that $f(x) = o(|x|)$ as $|x| \to \infty$. The way to prove this result is by a quantitative estimates on the coupling: There is a coupling for random walk starting at vertices distance 2 apart so that the probability of no coupling before exiting a ball of radius r decays like $\frac{c}{r}$.

12.2 Entropy

Conditional Expectation. Let $(\Omega, \mathcal{F}, \mathbb{P})$ be a probability space. For an integrable random variable X and a σ−algebra $\mathcal{G}$ recall that $\mathbb{E}[X|\mathcal{G}]$, called the conditional expectation of X w.r.t $\mathcal{G}$, is the unique (a.s.) $\mathcal{G}$-measurable random variable that satisfies $\mathbb{E}[\mathbb{1}_A X] = \mathbb{E}[\mathbb{1}_A \mathbb{E}[X|\mathcal{G}]]$ for every $A \in \mathcal{G}$.

Conditional expectation has the following two properties (see e.g. [Dur10, Chap. 5]) that we will need:

Proposition 12.9 (Tower Property). *For $\mathcal{G}_1 \subseteq \mathcal{G}_2$*

$$\mathbb{E}[\mathbb{E}[X|\mathcal{G}_2]|\mathcal{G}_1] = \mathbb{E}[X|\mathcal{G}_1].$$

Recall that a function $\phi : (a,b) \to \mathbb{R}$ is *convex* if for any $x, y \in (a,b)$ and $\alpha \in [0,1]$

$$\phi(\alpha x + (1-\alpha)y) \leq \alpha f(x) + (1-\alpha)f(y).$$

ϕ is *strictly convex* if equality holds above only in the (trivial) cases $\alpha = 0$ or $\alpha = 1$. For example, a function with non-negative second derivative is convex. A function with strictly positive second derivative is strictly convex. A function ϕ is *concave* (resp. *strictly concave*) if $-\phi$ is convex (resp. strictly convex).

Lemma 12.10 (Jensen's Inequality). *Let ϕ be a convex function. Assume that* $\mathbb{E}[|X|], \mathbb{E}[|\phi(X)|] < \infty$. *Then,* $\phi(\mathbb{E}[X|\mathcal{G}]) \leq \mathbb{E}[\phi(X)|\mathcal{G}]$. *Moreover, if ϕ is strictly convex, then* $\phi(\mathbb{E}[X]) = \mathbb{E}[\phi(X)]$ *implies that* $X = \mathbb{E}[X]$ *a.s. (i.e. X is a.s. constant).*

Entropy. All logarithms in this section are base 2. In addition we use the convention $0 \log 0 = 0$.

Let X be a discrete random variable. For any x, let $p(x)$ be the probability that $X = x$. The *entropy* of X is defined as

$$H(X) = \mathbb{E}[-\log p(X)] = -\sum_x p(x) \log p(x).$$

Let σ be a σ-algebra on Ω. For a discrete random variable X, we can define the conditional probability $p_\sigma(x) = \mathbb{P}(X = x|\sigma) = \mathbb{E}[\mathbb{1}_{\{X=x\}}|\sigma]$. Since a.s. $\sum_x p_\sigma(x) = 1$, we have a "random" entropy

$$H_\sigma(X) = -\sum_x p_\sigma(x) \log p_\sigma(x).$$

Taking expectation, we define

$$H(X|\sigma) = \mathbb{E}[H_\sigma(X)].$$

If Y is another discrete random variable we define $H(X|Y) = H(X|\sigma(Y))$ and $H(X,Y)$ is the entropy of the random variable (X,Y), i.e., $H(X,Y) = -\sum_{x,y} p(x,y) \log p(x,y)$, where $p(x,y) = \mathbb{P}(X = x, Y = y)$.

Proposition 12.11. *The following relations hold:*

1. $H(X|Y) = H(X,Y) - H(Y)$.
2. *If $\mathcal{G}_1 \subseteq \mathcal{G}_2$ then $H(X|\mathcal{G}_2) \leq H(X|\mathcal{G}_1)$.*
3. $H(X|\mathcal{G}) = H(X)$ *if and only if X is independent of $\mathcal{G}$.*

For more information on entropy see e.g., [CT06, Chap. 2].

Proof. Part 1: Since

$$P(X = x | \sigma(Y)) = \mathbb{E}[\mathbb{1}_{\{X=x\}} | \sigma(Y)] = \sum_y \mathbb{1}_{\{Y=y\}} \frac{P(X = x, Y = y)}{P(Y = y)}$$

we get that $H(X|Y) = -\sum_{x,y} P(X = x, Y = y) \log P(X = x | Y = y)$.
Therefore

$$H(X, Y) = -\sum_{x,y} P(X = x, Y = y) \log P(X = x, Y = y)$$

$$= -\sum_{x,y} P(X = x, Y = y) \log P(X = x | Y = y) - \sum_{x,y} P(X = x, Y = y) \log P(Y = y)$$

$$= H(X|Y) + H(Y).$$

Part 2: The function $\phi(x) = -x \log x$ is a concave function on $(0, 1)$. Thus, Jensen's inequality tells us that

$$H(X|\mathcal{G}_2) = \sum_x \mathbb{E}[\phi(\mathbb{E}[\mathbb{1}_{\{X=x\}}|\mathcal{G}_2])] = \sum_x \mathbb{E}\left[\mathbb{E}[\phi(\mathbb{E}[\mathbb{1}_{\{X=x\}}|\mathcal{G}_2])|\mathcal{G}_1]\right]$$

$$\leq \sum_x \mathbb{E}\left[\phi(\mathbb{E}[\mathbb{E}[\mathbb{1}_{\{X=x\}}|\mathcal{G}_2]|\mathcal{G}_1])\right] = \sum_x \mathbb{E}\left[\phi(\mathbb{E}[\mathbb{1}_{\{X=x\}}|\mathcal{G}_1])\right]$$

$$= H(X|\mathcal{G}_1).$$

Part 3: We use the "equality version" of Jensen's inequality. The function $\phi(x) = -x \log x$ is strictly concave on $(0, 1)$. Since $H(X|\sigma) \leq H(X)$, we have that equality holds iff $\mathbb{E}[\phi(\mathbb{E}[\mathbb{1}_{\{X=x\}}|\sigma])] = \phi(\mathbb{P}[X = x])$ for every x. Thus, with $Z = \mathbb{E}[\mathbb{1}_{\{X=x\}}|\sigma]$, we have that this holds iff $\mathbb{E}[\mathbb{1}_{\{X=x\}}|\sigma] = \mathbb{P}[X = x]$ for every x a.s. which is iff X is independent of σ.   □

We will also need the following convergence theorem.

Lemma 12.12. *Let $(\mathcal{F}_n)$ be an increasing sequence of σ-algebras, increasing to $\mathcal{F}_\infty$. Let (σ_n) be a decreasing sequence of σ-algebras, decreasing to $\sigma_\infty = \cap_n \sigma_n$. Then, for any finitely supported random variable X,*

$$\lim_{n\to\infty} H(X|\mathcal{F}_n) = H(X|\mathcal{F}_\infty),$$

and

$$\lim_{n\to\infty} H(X|\sigma_n) = H(X|\sigma_\infty).$$

The proof is classical; the first assertion is known as *Lévy's $0 - 1$ Law* or *Lévy's Forward Theorem*, the second assertion is just backward martingale convergence. Both proofs can be found for example in [Dur10, Chap. 5]

12.3 The Furstenberg-Poisson Boundary

The following is a summary of the main result of [KV83].

Let G be a finitely generated group, and let $\{X_n\}_{n\geq 0}$ be a random walk on G (with respect to a fixed finite symmetric generating set). Let $\sigma_n = \sigma(X_n, X_{n+1}, \dots,)$, and $\mathcal{T} = \cap_n \sigma_n$ the tail σ-algebra of the random walk.

Let U_n be the increments of the random walk, i.e., U_n is defined by the relation $X_n = X_{n-1} U_n$. This implies that $\{U_n\}_{n\geq 1}$ are i.i.d.

Another σ−algebra that can be defined is the *invariant* σ-algebra, denoted $\mathcal{I}$. Consider the space $\Omega \subset G^{\mathbb{N}}$ of infinite trajectories of nearest-neighbor (w.r.t to the generating set) walks in G. The shift operator $\theta : \Omega \to \Omega$ is defined by $\theta((\omega_0, \omega_1, \dots)) = (\omega_1, \omega_2, \dots)$. A set $A \subset \Omega$ is called invariant if $\theta^{-1}(A) = A$. The invariant σ-algebra $\mathcal{I}$ is defined to be the family of all invariant (measurable) sets.

It is clear from the definition that any invariant set is also an event in the tail σ-algebra $\mathcal{T}$.

Let us construct these two σ−algebras in an alternative way: For $\omega, \omega' \in \Omega$, denote $\omega \sim \omega'$ if there exist $k, N \in \mathbb{N}$ such that $\omega_{n+k} = \omega'_n$ for all $n > N$. The relation $\sim$ defines an equivalence relation on Ω and therefore we can define $\Gamma = \Omega/\sim$. The σ-algebra generated by the equivalence classes in Γ is exactly the invariant σ−algebra $\mathcal{I}$.

For every $x \in G$, there is a natural measure ν_x on Γ, namely the image of $\mathbb{P}_x$ (the original measure on Ω conditioned that $\omega_0 = x$) under the quotient map. This measures have the following properties:

- The space (Γ, ν) for $\nu = \nu_e$ (where e is the unit element in G) is isomorphic to the Poisson-Furstenberg boundary as a measure space with a G-action on the measure. For more details on the Poisson-Furstenberg boundary see [Fur63].
- Note that $\mathbb{E}_x[\nu_{X_1}] = \nu_x$, so ν is G-stationary.

Let $\mathcal{P}$ denote the Markov operator of the random walk; that is, for $f : G \to \mathbb{R}$ define $(\mathcal{P}f)(x) = \mathbb{E}_x[f(X_1)]$.

Definition 12.13. An *harmonic* function on G is a function $f : G \to \mathbb{R}$ such that $\mathcal{P}f = f$ (that is, $f(X_n)$ is a martingale).

Lemma 12.16 explains the connection between harmonic functions on G and $\mathcal{I}$.

Similarly, one can construct $\mathcal{T}$ from the following equivalence relation: given $\omega, \omega' \in \Omega$ denote $\omega \approx \omega'$ if there exists $n \in \mathbb{N}$ such that $\theta^n(\omega) = \theta^n(\omega')$. As before it is not hard to show that this is indeed an equivalence relation. As for the invariant σ-algebra the tail σ-algebra is exactly the σ-algebra generated by the equivalence classes in $\Omega/\approx$.

Definition 12.14. A *parabolic* function (sometimes called a space-time harmonic function) is a function $f : G \times \mathbb{N} \to \mathbb{R}$ such that $f(X_n, n)$ is a martingale; that is, a sequence of functions $f_n : G \to \mathbb{R}$ such that $\mathcal{P}f_{n+1} = f_n$.

The relation between parabolic functions and $\mathcal{T}$ is also explained in Lemma 12.16.

A note about triviality of σ-algebras: We have a large collection of different measures we can relate to when speaking of triviality of σ-algebras. We thus define

Definition 12.15. A $\sigma-$algebra $\mathcal{F}$ is called trivial if for every $A \in \mathcal{F}$ and every $x \in G$ we have that $\mathbb{P}_x(A) \in \{0, 1\}$.

Lemma 12.16. *The correspondence $f \mapsto \lim_{n\to\infty} f(X_n)$ is an invertible correspondence between bounded harmonic functions on G and bounded $\mathcal{I}$-measurable random variables, whose inverse is given by $L \mapsto f$ where $f(x) = \mathbb{E}_x[L]$. In particular, $\mathcal{I}$ is trivial if and only if any bounded harmonic function on G is constant.*

Also, there exists a similar correspondence between bounded $\mathcal{T}$-measurable random variables and bounded parabolic functions. Again, in particular $\mathcal{T}$ is trivial if and only if any bounded parabolic function on G is constant.

Proof. Let f be a bounded harmonic function on G. This implies that $f(X_n)$ is a bounded martingale, and thus converges a.s. to some limit L. Because $f(X_n)$ and $f(X_{n+1}) = f(X_n(\theta(\omega)))$ converge to the same limit, we have that $L(\omega) = L(\theta(\omega))$ $\mathbb{P}_x$-a.s. for every $x \in G$. Therefore L is bounded and measurable with respect to $\mathcal{I}$.

Conversely, if L is a $\mathcal{I}$-measurable bounded random variable, then the function $f(x) = \mathbb{E}_x[L]$ determines a bounded harmonic function, because

$$\mathcal{P}f(x) = \mathbb{E}_x[f(X_1)] = \mathbb{E}_x\left[\mathbb{E}_{X_1}[L(X_1, X_2, \ldots)]\right] = \mathbb{E}_x\left[\mathbb{E}_{X_1}[L \circ \theta(X_0, X_1, \ldots)]\right]$$
$$= \mathbb{E}_x[L \circ \theta(X_0, X_1, \ldots)] = \mathbb{E}_x[L(X_0, X_1, \ldots)] = f(x).$$

Since the correspondence implies that L is a.s. constant if and only if f is constant, it follows that if $\mathcal{I}$ is trivial then f can only be a constant function. Also, if any bounded harmonic function is constant, the for any $\mathcal{I}$-measurable function L, we have $\mathbb{E}_x[L] = \mathbb{E}_e[L]$ for any $x \in G$. Thus,

$$\mathbb{E}_x[L|U_1, \ldots, U_n] = \mathbb{E}_{X_n}[L] = \mathbb{E}_x[L],$$

so L is independent of $U_1, \ldots, U_n$ for any $n \in \mathbb{N}$. Since L is measurable with respect to $\sigma(X_0, U_1, \ldots) = \sigma_0$, by Kolmogorov's $0 - 1$ law L must be a constant a.s.

We conclude that $\mathcal{I}$ is trivial if and only if every bounded harmonic function is constant.

The proof for $\mathcal{T}$ is similar: If $f : G \times \mathbb{N} \to \mathbb{R}$ is a bounded parabolic function, then $f(X_n, n)$ is a bounded martingale, and in particular converges a.s. to a limit L. Because for every $m \in \mathbb{N}$ we have $L = \lim_{\substack{n\to\infty \\ n \geq m}} f(X_n, n)$ it follows that L is σ_m-measurable for every $m \in \mathbb{N}$. Thus, L is $\mathcal{T}$-measurable. For the other direction, if L is a bounded $\mathcal{T}$-measurable random variable, for any $n \in \mathbb{N}$ one can write $L = L_n \circ \theta^n$ for some random variable L_n. Define $f : G \times \mathbb{N} \to \mathbb{R}$ by $f(x, n) = \mathbb{E}_x[L_n]$. Then,

$$\mathbb{E}_x[L|X_0,\ldots,X_n] = \mathbb{E}_x[L_n \circ \theta^n(X_0,X_1,\ldots)|X_0,\ldots,X_n]$$
$$= \mathbb{E}_x[L_n(X_n,\ldots)|X_0,\ldots,X_n] = \mathbb{E}_{X_n}[L_n] = f(X_n,n).$$

So $f(X_n,n)$ is a martingale (with respect to the filtration $\sigma(X_0,\ldots,X_n)$); i.e. f is a parabolic function. Note that $f(X_n,n)$ converges a.s. to L.

Similarly to the case of $\mathcal{I}$, we can conclude that $\mathcal{T}$ is trivial if and only if every bounded parabolic function is constant.   □

Remark 12.17. The correspondence between harmonic functions and $\mathcal{I}$ measurable functions is the reason for the name *Poisson boundary*. In the unit disc of the complex plane, for every point in the disc there is an associated measure on the boundary of the disc, namely the exit measure of Brownian motion in the plane. Given a function on the boundary, integrating the function against this measure gives us an harmonic function in the disc.

The construction of (Γ, ν) above generalizes this phenomena. We constructed a set Γ of different "behaviors at infinity", along with an "exit measure" for every point in G, so that if we integrate a function on Γ (i.e. an invariant function) against these exit measures—we obtain an harmonic function on G. This idea was first introduced by Furstenberg. The construction via $\mathcal{I}$ is due to Kaimanovich and Vershik.

Since the property that all bounded harmonic functions are constant turns out to be important, let us name it:

Definition 12.18. A graph is called *Liouville* if any bounded harmonic function on it is constant.

Thus we have shown

Corollary 12.19.

$$\text{a graph is Liouville} \iff \text{the invariant } \sigma\text{-algebra } \mathcal{I} \text{ is trivial.} \qquad (12.1)$$

12.4   Entropy and the Tail σ-algebra

Recall the definition of conditional entropy $H(X_1,\ldots,X_k|\sigma) = \mathbb{E}\left[-\sum_x p_\sigma(x)\right.$ $\left.\log p_\sigma(x)\right]$, where $p_\sigma(x) = \mathbb{P}[X = x|\sigma] = \mathbb{E}[\mathbb{1}_{\{X=x\}}|\sigma]$.

Claim 12.20. *For any $k < n$, and any $m > 0$,*

1.

$$H(X_1,\ldots,X_k|X_n) = H(X_1,\ldots,X_k|X_n,\ldots,X_{n+m}) = H(X_1,\ldots,X_k|\sigma_n).$$

2.

$$H(X_1,\ldots,X_k,X_n) = kH(X_1) + H(X_{n-k}).$$

3.

$$H(X_n) - H(X_{n-1}) = H(X_1) - H(X_1|\sigma_n)$$

Proof. The first assertion follows from applying the Markov property at time n, and from the fact that the σ-algebras $\{\sigma(X_n,\ldots,X_{n+m})\}_{m\geq 0}$ increase to $\sigma_n = \sigma(X_n, X_{n+1},\ldots)$.

The second assertion follows by applying induction on k and from the fact that

$$H(X_1,\ldots,X_k,X_n) = H(X_2,\ldots,X_k,X_n|X_1) + H(X_1) = H(X_1,\ldots,X_{k-1},X_{n-1}) + H(X_1).$$

The last equality holds since G is a group, so conditioned on $X_1 = x$ we have that $(X_2,\ldots,X_k,X_n)$ has the same distribution as $(xX_1,\ldots,xX_{k-1},xX_{n-1})$.

The last assertion follows from the first ones by the following argument: Since

$$H(X_n)-H(X_{n-k}) = kH(X_1)-(H(X_1,\ldots,X_k,X_n)-H(X_n)) = kH(X_1)-H(X_1,\ldots,X_k|X_n),$$

and in particular

$$H(X_n) - H(X_{n-1}) = H(X_1) - H(X_1|X_n). \qquad \square$$

Using the last proof we can define the following:

Definition 12.21. The asymptotic entropy of a finitely generated group G with respect to a generator set S is

$$h(G) = \lim_{n\to\infty} \frac{H(X_n)}{n},$$

where X_n is a random walk on the group generated by the generator set S. Note that by Claim 12.20 part 3 the last limit exists almost surely. Indeed, the sequence $H(X_1|X_n)$ is increasing and therefore $H(X_n)-H(X_{n-1})$ is a non negative decreasing sequence. This implies that its limit exists and so the limit of its averages converge as well. However

$$h(G) = \lim_{n\to\infty} \frac{H(X_n)}{n} = \lim_{n\to\infty} \frac{H(X_n) - H(X_1)}{n}$$

$$= \lim_{n\to\infty} \frac{1}{n}\sum_{k=1}^{n} H(X_k) - H(X_{k-1}) = \lim_{n\to\infty} H(X_n) - H(X_{n-1})$$

(12.2)

and therefore the required limit exists.

Using all of the above we can deduce

Lemma 12.22. $h = 0$ *if and only if* T *is trivial.*

Proof. Using (12.2) for any k one can write

$$kh = \lim_{n\to\infty} \sum_{j=0}^{k-1} H(X_{n-j}) - H(X_{n-j-1}) = \lim_{n\to\infty} H(X_n) - H(X_{n-k})$$

$$= \lim_{n\to\infty} kH(X_1) - H(X_1,\ldots,X_k|X_n) = \lim_{n\to\infty} kH(X_1) - H(X_1,\ldots,X_k|\sigma_n)$$

$$= kH(X_1) - H(X_1,\ldots,X_k|T) = H(X_1,\ldots,X_k) - H(X_1,\ldots,X_k|T).$$

Consequently, if $h = 0$ then T is independent of any $(X_1,\ldots,X_k)$, and thus independent of any $(U_1,\ldots,U_k)$. This implies that T is trivial by the Kolmogorov $0-1$ law. On the other hand, if T is trivial, then $H(X_1|T) = H(X_1)$ and so $h = 0$. $\square$

12.5 Entropy and the Poisson Boundary

Summarizing the previous sections, so far we showed that

$$h(G) = 0 \qquad G \text{ is Liouville}$$

$$\Updownarrow \qquad\qquad\qquad \Updownarrow$$

$$T \text{ is trivial} \ \Rightarrow\ \mathcal{I} \text{ is trivial}$$

This shows that 0 asymptotic entropy implies the triviality of the Poisson boundary Γ, or equivalently the Liouville property. We will now show that triviality of $\mathcal{I}$ implies triviality of T and thus that a finitely generated group is Liouville if and only if $h(G) = 0$.

First an auxiliary lemma.

Lemma 12.23. *For any* $p \in (0,1)$ *there exists a constant* $c(p) > 0$ *such that for any* $N \geq 0$ *there exists a coupling* (B, B') *of binomial-*(N, p) *and binomial-*$(N+1, p)$ *random variables, such that* $\mathbf{P}(B \neq B') \leq c(p)N^{-1/2}$ *(*$\mathbf{P}$ *is the coupling measure).*

Proof. Let $X_1, X_2,\ldots, Y_1, Y_2,\ldots$, be i.i.d. Bernoulli-$p$ random variables and define $R_n = \sum_{j=1}^{n} X_j - Y_j$. So $\{R_n\}_{n\geq 0}$ is a lazy (symmetric) random walk on $\mathbb{Z}$, that stays put with probability $p^2 + (1-p)^2 = 1 - 2p(1-p)$. Let $T = \inf\{n \geq 1 : R_n = 1\}$. Now define

$$B_X(N) = \begin{cases} \sum_{j=1}^{N} X_j & , N \leq T \\ \sum_{j=1}^{T} X_j + \sum_{j=T+1}^{N} X_j & , N > T \end{cases}$$

and

$$B_Y(N) = \begin{cases} \sum_{j=1}^{N} X_j & , N \leq T \\ \sum_{j=1}^{T} Y_j + \sum_{j=T+1}^{N} X_j & , N > T \end{cases} .$$

Then by classical random walk estimates

$$\mathbf{P}(B_X(N) \neq B_Y(N) + 1) \leq \mathbf{P}(T \geq N) \leq \frac{c(p)}{N^{1/2}}.$$

Now, let (B_X, B_Y) be a coupling as above, and let ξ be a Bernoulli-p random variable independent of the binomial coupling. Set

$$(B, B') = \begin{cases} (B_X, B_Y + 1) & \xi = 1 \\ (B_X, B_X) & \xi = 0. \end{cases}$$

Then (B, B') is a coupling of binomial-(N, p) and binomial-$(N + 1, p)$ random variables, and

$$\mathbf{P}(B \neq B') = p\mathbf{P}(B_X \neq B_Y + 1) \leq \frac{pc(p)}{N^{1/2}}. \qquad \square$$

Lemma 12.24. *Let $0 < p < 1$ and let $\mathcal{Q}_p = p \cdot \mathrm{Id} + (1 - p)\mathcal{P}$; that is $\mathcal{Q}_p$ is the Markov operator of a lazy random walk on G that stays put with probability p. Then, any bounded parabolic function with respect to $\mathcal{Q}_p$ is an harmonic function. More specifically, the tail and invariant σ−algebras with respect to $\mathcal{Q}_p$ coincide.*

Proof. Let $\{L_n\}$ be a lazy random walk on G, that stays put with probability $p > 0$. L_n has the same distribution as X_{B_n}, where B_n is a binomial-(n, p) random variable, and X_n is a random walk on G (i.e. with Markov operator $\mathcal{P}$). Let f_n be a bounded parabolic function with respect to $\mathcal{Q}_p$, with uniform bound M. Fix m and take n large. Let (B, B') be a coupling of binomial-(n, p) and binomial-$(n + 1, p)$ random variables, so that $\mathbf{P}(B \neq B') \leq c(p)n^{-1/2}$, as in Lemma 12.23. We then have

$$|f_m(x) - f_{m+1}(x)| = |\mathbb{E}_x[f_{m+n+1}(L_{n+1}) - f_{m+n+1}(L_n)]|$$
$$= |\mathbb{E}_x \mathbf{E}[f_{m+n+1}(X_{B'}) - f_{m+n+1}(X_B)]|$$
$$\leq 2M\mathbf{P}(B \neq B') \leq 2Mc(p)n^{-1/2}.$$

Taking n to infinity, we have that $f_m(x) = f_{m+1}(x)$ for all x and all m. Thus, $f_n = f_0$ is an harmonic function. $\qquad \square$

Lemma 12.25. *Let $x \in G$. For any tail event $A \in \mathcal{T}$, there exists an invariant event $B \in \mathcal{I}$ such that $\mathbb{P}_x[A] = \mathbb{P}_x[B]$. In particular, $\mathcal{T}$ is $\mathbb{P}_x$-trivial if and only if $\mathcal{I}$ is $\mathbb{P}_x$-trivial.*

Proof. Let f_n be a bounded parabolic function. First we show that $f_{2n} = f_0$ and $f_{2n+1} = f_1$ for all $n \in \mathbb{N}$. Indeed, $(X_{2n})_{n \geq 0}$ is a lazy random walk on (a transitive subgraph of) G, with Markov operator $\mathcal{P}^2$, where $\mathcal{P}^2 = p \cdot \mathrm{id} + (1 - p)\mathcal{P}'$ with $\mathcal{P}'(x, y) = (1 - p)^{-1}\mathbb{P}_x(X_2 = y)\mathbb{1}_{\{y \neq x\}}$ and $p = \mathbb{P}_x(X_2 = x)$. The sequences $\{f_{2n}\}_{n \geq 0}$ and $\{f_{2n+1}\}_{n \geq 0}$ are bounded parabolic functions with respect to $\mathcal{P}^2$. By Lemma 12.24 we get that $f_{2n} = f_0$ and $f_{2n+1} = f_1$ for all $n \in \mathbb{N}$.

Next consider two cases:

Case 1: G is not bi-partite; that is, there exists $k \in \mathbb{N}$ such that $\mathbb{P}_x(X_{2k+1} = x) > 0$ (which is independent of x by transitivity).

Case 2: G is bi-partite; that is, $\mathbb{P}_x(X_{2k+1} = x) = 0$ for all $k \in \mathbb{N}$.

In the first case, fix k with the required property and let $q = \mathbb{P}_x(X_{2k+1} = x) > 0$. Consider the Markov operator

$$\mathcal{P}''(x, y) = (1 - q)^{-1}\mathbb{P}_x(X_{2k+1} = y)\mathbb{1}_{\{y \neq x\}}.$$

Since $\mathcal{P}^{2k+1} = q \cdot \mathrm{Id} + (1-q)\mathcal{P}''$ and $\mathcal{P}^{2k+1} f_{n+2k+1} = f_n$, we get that the sequence $\{f_{(2k+1)n}\}$ is parabolic with respect to the lazy walk $\mathcal{P}^{2k+1}$, and thus, $f_{2k+1} = f_0$. Since we already proved that $f_{2k+1} = f_1$ this proves that in the non-bi-partite case, $f_0 = f_1$ and therefore f_0 is harmonic.

For the second case, where G is bi-partite, $f_n = f_{(n \bmod 2)}$ is *not* harmonic in general. However, we have the following construction. Since G is bi-partite, we have that $V(G) = V^+ \cup V^-$, where $V^+ \cap V^- = \emptyset$, and for $\xi \in \{+, -\}$ if $x \in V^\xi$ then $X_{2n} \in V^\xi$ and $X_{2n+1} \in V^{-\xi}$ for all n, $\mathbb{P}_x$-a.s. We can define

$$g^{\pm}(x) = \begin{cases} f_0(x) & x \in V^\pm \\ f_1(x) & x \in V^\mp. \end{cases}$$

One can check that $\mathcal{P}g^{\pm} = g^{\pm}$, so $g^{\pm}$ are bounded harmonic functions.

To conclude the lemma we have the following: If $A \in \mathcal{T}$ is a tail event, then there exists a bounded parabolic function f_n such that $f_n(X_n)$ converges $\mathbb{P}_x$-a.s. to $\mathbb{1}_A$, as in Lemma 12.16. However, for $\xi \in \{+, -\}$ such that $x \in V^\xi$, and for g^ξ as above, we have that $g^\xi(X_{2n}) = f_0(X_{2n}) = f_{2n}(X_{2n})$ which converges $\mathbb{P}_x$-a.s. to $\mathbb{1}_A$. However, the limit satisfies $\lim_{n \to \infty} g^\xi(X_{2n}) = \lim_{n \to \infty} g^\xi(X_n) = \mathbb{1}_B$ $\mathbb{P}_x$-a.s. for some invariant event $B \in \mathcal{I}$, because g^ξ is bounded harmonic. Thus, $A = B$, $\mathbb{P}_x$-a.s.   □

12.6 Conclusion

We summarize what we have shown:

We associated a few notions with a graph, and specifically a Cayley graph (a graph generated from a finitely generated group, with a specified symmetric generating set). The tail and invariant σ-algebras, harmonic and parabolic functions, asymptotic entropy, Liouville property. We showed that

- A graph is Liouville if and only if the invariant σ-algebra is trivial, see Corollary 12.19.
- *On groups:* zero asymptotic entropy is equivalent to a trivial tail σ-algebra, Lemma 12.22.
- *On transitive graphs:* The tail σ-algebra is trivial if and only if the invariant σ-algebra is trivial Lemma 12.25.

Altogether, we have an entropic criterion for Liouville, or triviality of the Poisson-Furstenberg boundary:

Theorem 12.26 (Kaĭmanovich and Vershik, [KV83]). *The Poisson-Furstenberg boundary of a finitely generated group is trivial (i.e. the group is Liouville) if and only if the asymptotic entropy is 0.*

Remark 12.27. The main place we used the group structure is in the claim that

$$H(X_2, \ldots, X_k, X_n | X_1) = H(X_1, \ldots, X_{k-1}, X_{n-1}).$$

The equivalence between zero entropy and Liouville does not hold for general graphs. In fact, the above property is what fails in the example in the next section, where a Liouville graph is constructed with positive asymptotic entropy.

12.7 Algebraic Recurrence of Groups

In a joint work with Hilary Finucane and Romain Tessera we initiate the study of a notion somewhat related to Liouville property for groups.

Let G be a countable group, μ be a probability measure on G, $\zeta_i \sim \mu$ be i.i.d., and let $X_n = \zeta_1 \zeta_2 \cdots \zeta_n$. Then we call $(X_1, X_2, \ldots)$ a *μ-random walk on G*. If $supp(\mu)$ does not generate G, then the random walk is restricted to a subgroup of G. Thus, we are interested only in measures μ whose support generates G. We will also restrict our focus to *symmetric* measures; i.e. measures μ such that $\mu(g) = \mu(g^{-1})$ for all $g \in G$.

Definition 12.28. Let $(X_1, X_2, \ldots)$ be a μ-random walk on G, and let S_n denote the semigroup generated by $\{X_n, X_{n+1}, \ldots\}$. We say (G, μ) is *algebraically recurrent* (AR) if for all $n \in \mathbb{N}$, $S_n = G$ almost surely, and we call G AR if (G, μ) is AR for all symmetric measures μ with $\langle supp(\mu) \rangle = G$. It remains open whether (G, μ) algebraic recurrence is independent of μ.

By showing in particular that $\mathbb{Z}^d$ is AR and the free group on 8 generators is not AR, we see that the algebraic recurrent groups is a nontrivial class of groups. We still don't have a good understanding of AR. For example we believe that Liouville groups are AR but this is still open.

The use of a semigroup rather than a subgroup in the definition may seem unnatural, but in fact the property is trivial if defined in terms of subgroups rather than semigroups. To see this, let G_n denote the *group* generated by $\{X_n, X_{n+1}, \ldots\}$ and suppose $\langle supp(\mu) \rangle = G$. Then for each $g \in supp(\mu)$, $\mathbb{P}(g \notin \{\zeta_{n+1}, \zeta_{n+2}, \ldots\}) = 0$, but for all $i > n$, $\zeta_i = (X_{i-1})^{-1} X_i \in G_n$. Thus, $supp(\mu) \subset G_n$, and so $G_n = G$ almost surely. This argument in fact proves the following more general result:

Lemma 12.29. *If $X_i^{-1} \in S_n$ almost surely for all $i \geq n$, then $S_n = G$ almost surely.*

12.7.1 $\mathbb{Z}^d$ is Algebraic Recurrent

As a first example we prove that $\mathbb{Z}^d$ is algebraic recurrent, starting with the case $d = 1$.

Proof for $\mathbb{Z}$. By the symmetry of μ, almost surely for all n there will be $y^+, y^- \in S_n$ with $y^+ > 0$ and $y^- < 0$. Thus for each $i \geq n$, if $X_i > 0$ we can write $X_i y^- + (-y^- - 1)X_i = -X_i$. Since the LHS is in S_n, so is $-X_i$. Similarly, if $X_i < 0$ we have $(y^+ - 1)X_i + -X_i y^+ = -X_i$, and again the LHS is in S_n so $-X_i \in S_n$. Using Lemma 12.29, this shows that $\mathbb{Z}$ is AR. $\square$

Remark 12.30. It is not much more difficult to see that $(\mathbb{Z}^d, \mu_d)$ is AR, when μ_d is uniform over the standard generating set. Indeed, after finitely many steps, the random walk will visit d linearly independent points, generating the intersection of a full-dimension lattice with a cone. Eventually, the random walk will visit a point x that is in the opposite cone, and by adding arbitrarily large multiples of x, the entire lattice is in S_n. Since there are only finitely many cosets of the lattice, the random walk eventually visits each coset, showing that all of $\mathbb{Z}^d$ is in S_n.

However, this proof does not extend to arbitrary symmetric generating measures on $\mathbb{Z}^d$. For example, in $\mathbb{Z}^2$, μ could have a very heavy tail along the line $x = y$ and a very small weight along the line $x = -y$, so that there are cones that the random walk has non-zero probability never to intersect. So for the general case, a more subtle proof is needed.

The proof that $\mathbb{Z}^d$ is AR, uses the following simple fact: To prove that (G, μ) is AR, it suffices to show that the trace of a μ-random walk on G is almost surely not contained in any proper subsemigroup of G. This is useful for proving that $\mathbb{Z}^d$ is AR because we can easily describe the subsemigroups of it.

Lemma 12.31. *Every proper subsemigroup of $\mathbb{Z}^d$ is contained either in a proper subgroup of $\mathbb{Z}^d$ or in a half-space of $\mathbb{Z}^d$.*

Proof. Let S be a subgroup of Z^d. If 0 is not in the convex hull of S, then there is a halfspace containing S. Otherwise, by Caratheodory's theorem, there are points $x_1, \ldots, x_{d+1}$ and positive numbers $t_1, \ldots, t_{d+1}$ such that $\sum t_i x_i = 0$ and $x_1, \ldots, x_d$ are linearly independent. Thus, x_{d+1} is written as a linear combination of $x_1, \ldots x_d$, using only negative coefficients. This allows us to generate arbitrary linear combinations of $x_1, \ldots, x_d$, so the group H generated by $x_1, \ldots, x_d$ is contained in S. Let $\bar{S}$ denote the projection of S to $\mathbb{Z}^d / H$. Because $\mathbb{Z}^d / H$ is torsion, $\bar{S}$ is a subgroup. If $\bar{S} = \mathbb{Z}^d / H$, then $S = \mathbb{Z}^d$. Otherwise, S is contained in a proper subgroup of $\mathbb{Z}^d$. □

Theorem 12.32. $\mathbb{Z}^d$ *is AR for all $d \geq 1$.*

Proof. The proof follows by induction on the dimension. The base case $d = 1$ was established before. Let μ be a symmetric measure on $\mathbb{Z}^d$ with $\langle supp(\mu) \rangle = \mathbb{Z}^d$. We would like to show that almost surely, S_n is not contained in any subgroup of $\mathbb{Z}^d$, or in any halfspace of $\mathbb{Z}^d$. By the proof of Lemma 12.29, $G_n = G$ for all n almost surely, so $\{X_n, X_{n+1}, \ldots\}$ is not contained in any proper subgroup of G almost surely, so S_n is not contained in any proper subgroup of G almost surely.

To see that S_n is not contained in any halfspace almost surely, consider the radial projection of the random walk to the sphere S^{n-1} in $\mathbb{R}^n$. By the compactness of the sphere, there is a non-empty closed subset $A \subset S^{n-1}$ of accumulation points of the projected random walk.

Next we show that A is a deterministic set, i.e., there exists a set $T \subset S^{n-1}$ such that $\mathbb{P}(A = T) = 1$. In addition we will show that $A = -A$ almost surely. The symmetry of A follows from the symmetry of μ. To show that A is deterministic, we use the Liouville property of $\mathbb{Z}^d$: By Kolmogorov 0-1 law every event depending only on the tail of a random walk on $\mathbb{Z}^d$ has probability 0 or 1. Because the definition of A depends only on the tail of a random walk on $\mathbb{Z}^d$, any event depending only on A has probability 0 or 1. However the only probability measure on closed sets that satisfies this property is the delta measure; in other words, there is one closed set T such that $P(A = T) = 1$.

Using the last claim, there is a pair of points $x, -x$ in S^{n-1} that are almost surely accumulation points of the projected random walk. Thus, the random walk will almost surely won't be contained in any halfspace that does not contain x and $-x$. Let P denote the $d - 1$ dimensional subspace orthogonal to x. If the random walk is contained in halfspace containing x and $-x$, then its projection to P must be contained in a halfspace. But the projection of the random walk to P is a $d - 1$ dimensional random walk, and so is almost surely not contained in any halfspace by induction. □

Open problem 12.33. *Is it possible to couple two simple random walks on $\mathbb{Z}^3$ or $\mathbb{Z}^4$, staring distance 10 apart, so that with positive probability their paths will not intersect?*

Chapter 13
Nonamenable Liouville Graphs

In this section we present an example of a bounded degree graph with a positive Cheeger constant (i.e. nonamenable graph) which is Liouville, that is, it admits no non constant bounded harmonic functions. This example shows that the theorem proved in Sect. 12 cannot be extended to general graphs.

A much more complicated example of bounded geometry simply connected Riemannian manifold with these properties was constructed in [BC96]. Nonamenable Cayley graphs are not Liouville, see [KV83], since the question "does nonamenability implies non Liouville for general bounded degree graphs?" keeps coming up and the example below is transparent (unlike [BC96]), we decided to write it down. The last subsection contains related conjectures.

The basic idea of the construction is to start with a binary tree and add edges to it, in order "to collapse the different directions simple random walk (SRW) can escape to infinity". This will be done using *expanders*, which is a family of finite d-regular graphs, G_n, growing in size, for which the isoperimetric condition holds with the same constant for any set of at most half the size of G_n, for all n. For background on expanders see e.g. [HLW06].

Once the idea of the construction is clear a simple non random version can be constructed. We believe the analysis and the example might be of interest e.g. for models in genetics, when information mixes along generations.

13.1   The Example

Fix some 3-regular expander $\{G_n\}$ with $|G_n| = 2^n$. On the vertices of the nth level of the binary tree place the graph G_n.

Denote the resulting graph by G. Since the binary tree is nonamenable and we only added edges in order to create G it follows that G is also nonamenable. For the Liouville property, examine L_n, the nth level of the tree. Consider it as a probability space (with the uniform probability, i.e., the probability of each point is 2^{-n}). For

I. Benjamini, *Coarse 3Geometry and Randomness*, Lecture Notes in Mathematics 2100,   121
DOI 10.1007/978-3-319-02576-6_13,
© Springer International Publishing Switzerland 2013

$x \in L_n$ and $y \in L_{n+1}$ denote the probability that a random walk starting from x will first hit L_{n+1} in y by $p(x, y)$. The corresponding operator will be denoted by $T = T_n$, i.e. for every $\phi : L_n \to \mathbb{R}$ let $T\phi : L_{n+1} \to \mathbb{R}$ be defined by

$$(T\phi)(y) = \sum_{x \in L_n} \phi(x)p(x, y), \quad \forall y \in L_{n+1}.$$

Next, denote by $\|T\|_{p \to q}$ the norm of T as an operator from $L^p(L_n) \to L^q(L_{n+1})$. A simple calculation shows that $\|T\|_{1 \to 1} \le 1$. Furthermore, $\|T\|_{\infty \to \infty}$ is also at most 1. To see this write

$$\|T\phi\|_\infty = \max_{y \in L_{n+1}} \left| \sum_{x \in L_n} \phi(x)p(x, y) \right| \le \max_{y \in L_{n+1}} \sum_{x \in L_n} |\phi(x)| \, p(x, y) \le \|\phi\|_\infty \max_{y \in L_{n+1}} \sum_{x \in L_n} p(x, y)$$

so

$$\|T\|_{\infty \to \infty} \le \max_{y \in L_{n+1}} \sum_{x} p(x, y).$$

Now by time reversal, $\sum_x p(x, y)$ is the same as the expected number of times a random walk starting from y visits L_n before exiting the ball $B_n := \cup_{k \le n} L_k$. Here we need to define the time the random walker exits B_n only after the first step. This expectation is easy to calculate. Denote the number of visits by N. There is probability $\frac{1}{6}$ that the first step of the walker goes to B_n, which is necessary for $N \ne 0$. Afterwards each time our walker reaches L_n there is probability $\frac{1}{3}$ for it to exit B_n. Hence the number of visits after first entering B_n is a geometric variable with mean 3, and we get $\mathbb{E}[N] = \frac{1}{6} \cdot 3 = \frac{1}{2}$ so $\|T\|_{\infty \to \infty} \le \frac{1}{2} \le 1$. By the Riesz-Thorin interpolation theorem we get

$$\|T\|_{2 \to 2} \le 1.$$

Let now $u \in L_k$ be some vertex and for $n \ge k$ denote by μ_n the harmonic measure on L_n for SRW starting at u. Let $f_n = \mu_n - 2^{-n}$. In fact, f_n is simply applying T repeatedly to $\mathbb{1}_u - 2^{-k}$ where $\mathbb{1}_u$ is a Kronecker δ at u. Because T sends probability measures to probability measures we always get that $\sum_{x \in L_n} f_n(x) = 0$. Because of the probabilistic interpretation of T we can examine the first step and only then apply T. For the first step there is probability $\frac{1}{3}$ to go to L_{n+1}, probability $\frac{1}{2}$ to stay in L_n and probability $\frac{1}{6}$ to go back to L_{n-1}. Denoting these operators (normalized to have norm 1) by S_1, S_2 and S_3 we get

$$(T_n f) = \frac{1}{3} S_1 f + \frac{1}{2} T_n S_2 f + \frac{1}{6} T_n T_{n-1} S_3 f.$$

The same reasoning as above shows that $\|S_i\|_{2 \to 2} \le 1$ and further, because S_2 is the operator corresponding to a random walk on an expander, and because the average of f is zero we get $||S_2 f||_2 \leq (1-\lambda) f$ where λ is the spectral gap of the expander we put on the levels. In fact, let's define λ as the infimum of these spectral gaps over all levels so we do not need to define λ_n etc. Combining all the above we get

$$||Tf||_2 \leq (1 - \tfrac{1}{2}\lambda)||f||_2.$$

Thus $||f_n||_2 \to 0$ as $n \to \infty$.

To finish the proof we use the fact that $||T||_{\infty \to \infty} \leq 1$ so $||f_n||_\infty \leq C$. Hence we also get $||f_n||_1 \to 0$ as $n \to \infty$. To show that G is Liouville, use Poisson's formula for harmonic function h with respect to level n large.

$$|h(u)-h(v)| = \left| \sum_{w \in T_n} h(w)(\mu_n^u(w) - \mu_n^v(w)) \right| \leq 2 \max_{x \in G} |h(x)| ||\mu_n^u(w) - \mu^v(w)||_1 \xrightarrow[n]{} 0,$$

since h is bounded. Here again μ_n^u and μ_n^v denote the harmonic measures on level n for SRW starting at u and v respectively. $\square$

13.2   Conjectures and Questions

A first natural question coming to mind is: What happens if we replace the binary tree by another nonamenable graph, on which we add expanders on all spheres around a fixed vertex? We were able to show that there is such graph which is not Liouville. The construction, very roughly, starts with an unbalanced tree (i.e. a binary tree where the father is connected to its left child by a regular edge, but to its right child with a double edge). We then utilized the fact that the harmonic measure is highly non-uniform on the levels and added a graph which is an expander with respect to the uniform measure but not with respect to the harmonic measure. We skip all details.

Maybe a stronger spectral requirement will imply existence of non constant bounded harmonic functions: An infinite d-regular graph G is called *Ramanujan*, if the spectral radius of the Markov operator (acting on l^2 of the vertices) equals $2\frac{\sqrt{d-1}}{d}$ which is what it is for the d-regular tree. When G is connected, this spectral radius can be expressed as the limiting exponent of the probabilities of return of a random walk, that is, $\lim_{n \to \infty} \sqrt[n]{p_n(v,v)}$.

Conjecture 13.1. Infinite Ramanujan graphs are not Liouville.

See [GKN12] for partial affirmative results on the way to proving the conjecture.

For general graphs, Liouville property is unstable under quasi-isometries, see [Lyo87]. For a simpler example see [BR11] and [BS96a].

It will be useful to prove the following conjecture:

Conjecture 13.2. A bounded degree graph G, which is rough isometric to a nonamenable Cayley graph, is not Liouville.

The last conjecture will be useful in pushing the heuristic regarding an attack on the old conjecture mentioned above for Cayley graphs by adding more edges in a rough isometric manner to get a graph closely imitating the example above.

Another open problem in this direction is the following:

Open problem 13.3. *Assume $\mathbb{Z}$ acts on G by isometries, $H = G/\mathbb{Z}$ is Liouville and simple random walk on G visits every translation of the fundamental domain H infinitely often a.s. Is G Liouville?*

In his proof of Gromov's theorem, Kleiner proved that every Cayley graph admits a non-constant Lipschitz harmonic function, see [Kle10]. This cannot be true for a general graph because some graphs admits no non-constant harmonic functions at all. For example, the half-line, or a half ladder which admits only an exponentially growing harmonic functions other then the constants. Uri Bader asked which general graphs admit a non-constant Lipschitz harmonic function? Ori Gurel-Gurevich observed e.g. that every transient graph or any graph with more than one end admits a non-constant Lipschitz harmonic function.

References

[AB87]   M. Aizenman, D.J. Barsky, Sharpness of the phase transition in percolation models. Comm. Math. Phys. **108**(3), 489–526 (1987)

[ABC+91]   J.M. Alonso, T. Brady, D. Cooper, V. Ferlini, M. Lustig, M. Mihalik, M. Shapiro, H. Short, Notes on word hyperbolic groups, in *Group Theory from a Geometrical Viewpoint*, Trieste, 1990, ed. by H. Short (World Scientific Publishing, River Edge, 1991), pp. 3–63

[ABS04]   N. Alon, I. Benjamini, A. Stacey, Percolation on finite graphs and isoperimetric inequalities. Ann. Probab. **32**(3A), 1727–1745 (2004)

[AL07]   D. Aldous, R. Lyons, Processes on unimodular random networks. Electron. J. Probab. **12**(54), 1454–1508 (2007)

[Anc88]   A. Ancona, Positive harmonic functions and hyperbolicity, in *Potential Theory— Surveys and Problems*, Prague, 1987. Lecture Notes in Mathematics, vol. 1344 (Springer, Berlin, 1988), pp. 1–23

[Ang03]   O. Angel, Growth and percolation on the uniform infinite planar triangulation. Geom. Funct. Anal. **13**(5), 935–974 (2003)

[AS03]   O. Angel, O. Schramm, Uniform infinite planar triangulations. Commun. Math. Phys. **241**(2–3), 191–213 (2003)

[Bab97]   L. Babai, The growth rate of vertex-transitive planar graphs, in *Proceedings of the Eighth Annual ACM-SIAM Symposium on Discrete Algorithms*, New Orleans, 1997 (ACM, New York, 1997), pp. 564–573

[BC96]   I. Benjamini, J. Cao, Examples of simply-connected Liouville manifolds with positive spectrum. Differ. Geom. Appl. **6**(1), 31–50 (1996)

[BC11]   I. Benjamini, N. Curien, On limits of graphs sphere packed in Euclidean space and applications. Europ. J. Combin. **32**(7), 975–984 (2011)

[BC12]   I. Benjamini, N. Curien, Ergodic theory on stationary random graphs. Electron. J. Probab. **17**(93), 20 (2012)

[BC13]   I. Benjamini, N. Curien, Simple random walk on the uniform infinite planar quadrangulation: subdiffusivity via pioneer points. Geom. Funct. Anal. **23**(2), 501–531 (2013)

[BDCGS12] R. Bauerschmidt, H. Duminil-Copin, J. Goodman, G. Slade, Lectures on self-avoiding walks. in *Probability and Statistical Physics in Two and More Dimensions, Clay Mathematics Institute Proceedings*, vol. 15, ed. by D. Ellwood, C.M. Newman, V. Sidoravicius, W. Werner (Cambridge, MA, 2012), pp. 395–476

[BDCKY11] I. Benjamini, H. Duminil-Copin, G. Kozma, A. Yadin, Disorder, entropy and harmonic functions. arXiv preprint arXiv:1111.4853 (2011)

I. Benjamini, *Coarse 3Geometry and Randomness*, Lecture Notes in Mathematics 2100,   125
DOI 10.1007/978-3-319-02576-6,
© Springer International Publishing Switzerland 2013

[BFT12]   I. Benjamini, H. Finucane, R. Tessera, On the scaling limit of finite vertex transitive graphs with large diameter. arXiv preprint arXiv:1203.5624 Combinatorica (2014)

[BGT12]   E. Breuillard, B. Green, T. Tao, The structure of approximate groups. Publ. Math. IÍHÉS **116**(1), 115–221 (2012)

[BH05]   I. Benjamini, C. Hoffman, ω-periodic graphs. Electron. J. Combin. **12**, Research Paper 46, 12 pp. (2005) (electronic)

[Bis04]   M. Biskup, On the scaling of the chemical distance in long-range percolation models. Ann. Probab. **32**(4), 2938–2977 (2004)

[BK89]   R.M. Burton, M. Keane, Density and uniqueness in percolation. Commun. Math. Phys. **121**(3), 501–505 (1989)

[BK05]   I. Benjamini, G. Kozma, A resistance bound via an isoperimetric inequality. Combinatorica **25**(6), 645–650 (2005)

[BK13]   I. Benjamini, G. Kozma, Uniqueness of percolation on products with z. Lat. Am. J. Probab. Math. Stat. **10**, 15–25 (2013)

[BL90]   B. Bollobás, I. Leader, An isoperimetric inequality on the discrete torus. SIAM J. Discrete Math. **3**(1), 32–37 (1990)

[BLPS99a]   I. Benjamini, R. Lyons, Y. Peres, O. Schramm, Group-invariant percolation on graphs. Geom. Funct. Anal. **9**(1), 29–66 (1999)

[BLPS99b]   I. Benjamini, R. Lyons, Y. Peres, O. Schramm, Critical percolation on any nonamenable group has no infinite clusters. Ann. Probab. **27**(3), 1347–1356 (1999)

[BLPS01]   I. Benjamini, R. Lyons, Y. Peres, O. Schramm, Uniform spanning forests. Ann. Probab. **29**(1), 1–65 (2001)

[BLS99]   I. Benjamini, R. Lyons, O. Schramm, Percolation perturbations in potential theory and random walks, in *Random Walks and Discrete Potential Theory*, Cortona, 1997. Symposia Mathematica XXXIX (Cambridge University Press, Cambridge, 1999), pp. 56–84

[BNP11]   I. Benjamini, A. Nachmias, Y. Peres, Is the critical percolation probability local? Probab. Theory Relat. Fields **149**(1–2), 261–269 (2011)

[Bow95]   B.H. Bowditch, A short proof that a subquadratic isoperimetric inequality implies a linear one. Mich. Math. J. **42**(1), 103–107 (1995)

[BP11]   I. Benjamini, P. Papasoglu, Growth and isoperimetric profile of planar graphs. Proc. Am. Math. Soc. **139**(11), 4105–4111 (2011)

[BPP98]   I. Benjamini, R. Pemantle, Y. Peres, Unpredictable paths and percolation. Ann. Probab. **26**(3), 1198–1211 (1998)

[BR11]   I. Benjamini, D. Revelle, Instability of set recurrence and Green's function on groups with the Liouville property. Potential Anal. **34**(2), 199–206 (2011)

[BS92]   L. Babai, M. Szegedy, Local expansion of symmetrical graphs. Comb. Probab. Comput. **1**(1), 1–11 (1992)

[BS96a]   I. Benjamini, O. Schramm, Harmonic functions on planar and almost planar graphs and manifolds, via circle packings. Invent. Math. **126**(3), 565–587 (1996)

[BS96b]   I. Benjamini, O. Schramm, Percolation beyond $\mathbf{Z}^d$, many questions and a few answers. Electron. Commun. Probab. **1**(8), 71–82 (1996) (electronic)

[BS00]   M. Bonk, O. Schramm, Embeddings of Gromov hyperbolic spaces. Geom. Funct. Anal. **10**(2), 266–306 (2000)

[BS01a]   I. Benjamini, O. Schramm, Percolation in the hyperbolic plane. J. Am. Math. Soc. **14**(2), 487–507 (2001) (electronic)

[BS01b]   I. Benjamini, O. Schramm, Recurrence of distributional limits of finite planar graphs. Electron. J. Probab. **6**(23), 13 pp. (2001) (electronic)

[BS04]   I. Benjamini, O. Schramm, Pinched exponential volume growth implies an infinite dimensional isoperimetric inequality, in *Geometric Aspects of Functional Analysis*. Lecture Notes in Mathematics, vol. 1850 (Springer, Berlin, 2004), pp. 73–76

[BS09]   I. Benjamini, O. Schramm, Lack of sphere packing of graphs via non-linear potential theory. arXiv preprint arXiv:0910.3071 (2009)

[BST12]   I. Benjamini, O. Schramm, Á. Timár, On the separation profile of infinite graphs. Groups Geom. Dyn. **6**(4), 639–658 (2012)

[CFKP97]   J.W. Cannon, W.J. Floyd, R. Kenyon, W.R. Parry, Hyperbolic geometry, in *Flavors of Geometry*. Mathematical Sciences Research Institute Publications, vol. 31 (Cambridge University Press, Cambridge, 1997), pp. 59–115

[CMM12]   N. Curien, L. Ménard, G. Miermont, A view from infinity of the uniform infinite planar quadrangulation. arXiv preprint arXiv:1201.1052 (2012)

[CT06]   T.M. Cover, J.A. Thomas, *Elements of Information Theory*, 2nd edn. (Wiley-Interscience, Hoboken, 2006)

[Dek91]   F.M. Dekking, Branching processes that grow faster than binary splitting. Am. Math. Mon. **98**(8), 728–731 (1991)

[DS13]   J. Ding, A. Sly, Distances in critical long range percolation. arXiv preprint arXiv:1303.3995 (2013)

[dup]   B. Duplantier, S. Sheffield, Liouville quantum gravity and KPZ. Invent. Math. **185**(2), 333–393 (2011)

[Dur10]   R. Durrett, *Probability: Theory and Examples*. Cambridge Series in Statistical and Probabilistic Mathematics, 4th edn. (Cambridge University Press, Cambridge, 2010)

[EFW07]   A. Eskin, D. Fisher, K. Whyte, Quasi-isometries and rigidity of solvable groups. Pure Appl. Math. Q. **3**(4, part 1), 927–947 (2007)

[Ers03]   A. Erschler, On drift and entropy growth for random walks on groups. Ann. Probab. **31**(3), 1193–1204 (2003)

[FI79]   H. Fleischner, W. Imrich, Transitive planar graphs. Math. Slovaca **29**(2), 97–106 (1979)

[Fur63]   H. Furstenberg, Noncommuting random products. Trans. Am. Math. Soc. **108**, 377–428 (1963)

[Gal11]   J.F.L. Gall, Uniqueness and universality of the Brownian map. Arxiv preprint arXiv:1105.4842 (2011)

[GG13]   O. Gurel-Gurevich, A. Nachmias, Recurrence of planar graph limits. Ann. Math. **177**, 761–781 (2013)

[GH90]   É. Ghys, A. Haefliger, Groupes de torsion, in *Sur les groupes hyperboliques d'après Mikhael Gromov*, Bern, 1988. Progress in Mathematics, vol. 83 (Birkhäuser, Boston, 1990), pp. 215–226

[GKN12]   R. Grigorchuk, V.A. Kaimanovich, T. Nagnibeda, Ergodic properties of boundary actions and the Nielsen-Schreier theory. Adv. Math. **230**(3), 1340–1380 (2012)

[GL13]   G. Grimmett, Z. Li, Counting self-avoiding walks. arXiv:1304.7216 (2013)

[GM02]   A. Gamburd, E. Makover, On the genus of a random Riemann surface, in *Complex Manifolds and Hyperbolic Geometry*, Guanajuato, 2001. Contemporary Mathematics, vol. 311 (American Mathematical Society, Providence, 2002), pp. 133–140

[GM06]   N. Gantert, S. Müller, The critical branching Markov chain is transient. Markov Process. Relat. Fields **12**(4), 805–814 (2006)

[GN90]   G.R. Grimmett, C.M. Newman, Percolation in $\infty + 1$ dimensions, in *Disorder in Physical Systems* (Oxford Science Publication/Oxford University Press, New York, 1990), pp. 167–190

[GPKS06]   M. Gromov, P. Pansu, M. Katz, S. Semmes, *Metric Structures for Riemannian and non-Riemannian Spaces*, vol. 152 (Birkhäuser, Boston, 2006)

[GPY11]   L. Guth, H. Parlier, R. Young, Pants decompositions of random surfaces. Geom. Funct. Anal. **21**(5), 1069–1090 (2011)

[Gri99]   G. Grimmett, *Percolation*. Grundlehren der Mathematischen Wissenschaften [Fundamental Principles of Mathematical Sciences], vol. 321, 2nd edn. (Springer, Berlin, 1999)

[Häg97]   O. Häggström, Infinite clusters in dependent automorphism invariant percolation on trees. Ann. Probab. **25**(3), 1423–1436 (1997)

[Häg11]   O. Häggström, Two badly behaved percolation processes on a nonunimodular graph. J. Theor. Probab. **24**, 1–16 (2011)

[HLW06]   S. Hoory, N. Linial, A. Wigderson, Expander graphs and their applications. Bull.
          Am. Math. Soc. (N.S.) **43**(4), 439–561 (2006) (electronic)
[HM09]   O. Häggström, P. Mester, Some two-dimensional finite energy percolation processes.
          Electron. Commun. Probab. **14**, 42–54 (2009)
[Hou11]   C. Houdayer, Invariant percolation and measured theory of nonamenable groups.
          Arxiv preprint arXiv:1106.5337 (2011)
[HP99]   O. Häggström, Y. Peres, Monotonicity of uniqueness for percolation on Cayley
          graphs: all infinite clusters are born simultaneously. Probab. Theory Relat. Fields
          113(2), 273–285 (1999)
[HPS99]   O. Häggström, Y. Peres, R.H. Schonmann, Percolation on transitive graphs as a
          coalescent process: relentless merging followed by simultaneous uniqueness, in
          Perplexing Problems in Probability. Progress in Probability, vol. 44 (Birkhäuser,
          Boston, 1999), pp. 69–90
[HS90]   T. Hara, G. Slade, Mean-field critical behaviour for percolation in high dimensions.
          Commun. Math. Phys. **128**(2), 333–391 (1990)
[HS95]   Z.-X. He, O. Schramm, Hyperbolic and parabolic packings. Discrete Comput. Geom.
          14(2), 123–149 (1995)
[Imr75]   W. Imrich, On Whitney's theorem on the unique embeddability of 3-connected planar
          graphs, in *Recent Advances in Graph Theory (Proceedings of Second Czechoslovak
          Symposium)*, Prague, 1974 (Academia, Prague, 1975), pp. 303–306 (loose errata)
[Jár03]   A.A. Járai, Invasion percolation and the incipient infinite cluster in 2D. Commun.
          Math. Phys. **236**(2), 311–334 (2003)
[Kes90]   H. Kesten, Asymptotics in high dimensions for percolation. In *Disorder in Physical
          Systems* (Oxford Science Publication/Oxford University Press, New York, 1990),
          pp. 219–240
[Kle10]   B. Kleiner, A new proof of Gromov's theorem on groups of polynomial growth.
          J. Am. Math. Soc. **23**(3), 815–829 (2010)
[KN11]   G. Kozma, A. Nachmias, Arm exponents in high dimensional percolation. J. Am.
          Math. Soc. **24**(2), 375–409 (2011)
[Koz07]   G. Kozma, Percolation, perimetry, planarity. Rev. Mat. Iberoam. **23**(2), 671–676
          (2007)
[Kri04]   M.A. Krikun, A uniformly distributed infinite planar triangulation and a related
          branching process. Zap. Nauchn. Sem. S.-Peterburg. Otdel. Mat. Inst. Steklov.
          (POMI) **307**, 141–174 (2004); Teor. Predst. Din. Sist. Komb. i Algoritm. Metody.
          10, 282–283 (2004)
[KV83]   V.A. Kaĭmanovich, A.M. Vershik, Random walks on discrete groups: boundary and
          entropy. Ann. Probab. **11**(3), 457–490 (1983)
[Laa00]   T.J. Laakso, Ahlfors Q-regular spaces with arbitrary $Q > 1$ admitting weak Poincaré
          inequality. Geom. Funct. Anal. **10**(1), 111–123 (2000)
[LL10]   G.F. Lawler, V. Limic, *Random walk: A Modern Introduction*. Cambridge Studies in
          Advanced Mathematics, vol. 123 (Cambridge University Press, Cambridge, 2010)
[LP09]   J.R. Lee, Y. Peres, Harmonic maps on amenable groups and a diffusive lower bound
          for random walks. Arxiv preprint arXiv:0911.0274 (2009)
[LPP95]   R. Lyons, R. Pemantle, Y. Peres, Ergodic theory on Galton-Watson trees: speed of
          random walk and dimension of harmonic measure. Ergodic Theory Dyn. Syst. **15**(3),
          593–619 (1995)
[LS99]   R. Lyons, O. Schramm, Indistinguishability of percolation clusters. Ann. Probab.
          27(4), 1809–1836 (1999)
[LSS97]   T.M. Liggett, R.H. Schonmann, A.M. Stacey, Domination by product measures. Ann.
          Probab. **25**(1), 71–95 (1997)
[LT79]   R.J. Lipton, R.E. Tarjan, A separator theorem for planar graphs. SIAM J. Appl. Math.
          36(2), 177–189 (1979)
[Lyo87]   T. Lyons, Instability of the Liouville property for quasi-isometric Riemannian
          manifolds and reversible Markov chains. J. Differ. Geom. **26**(1), 33–66 (1987)

[Lyo09]   R. Lyons, Y. Peres, *Probability on Trees and Networks* (2009). (in preparation)

[Mei08]   J. Meier, *Groups, Graphs and Trees: An Introduction to the Geometry of Infinite Groups*. London Mathematical Society Student Texts, vol. 73 (Cambridge University Press, Cambridge, 2008)

[Mie13]   G. Miermont, The Brownian map is the scaling limit of uniform random plane quadrangulations. Acta Math. **210**(2), 319–401 (2013)

[Moh88]   B. Mohar, Isoperimetric inequalities, growth, and the spectrum of graphs. Linear Algebra Appl. **103**, 119–131 (1988)

[MP01]   R. Muchnik, I. Pak, Percolation on Grigorchuk groups. Commun. Algebra **29**(2), 661–671 (2001)

[MS93]   N. Madras, G. Slade, *The Self-Avoiding Walk*. Probability and its Applications (Birkhäuser, Boston, 1993)

[MTTV98] G.L. Miller, S.-H. Teng, W. Thurston, S.A. Vavasis, Geometric separators for finite-element meshes. SIAM J. Sci. Comput. **19**(2), 364–386 (1998)

[Nek05]   V. Nekrashevych, *Self-Similar Groups*. Mathematical Surveys and Monographs, vol. 117 (American Mathematical Society, Providence, 2005)

[NP11]   V. Nekrashevych, G. Pete, Scale-invariant groups. Groups Geom. Dyn. **5**(1), 139–167 (2011)

[Pan]   P. Pansu, Packing a space into an other. (in preparation)

[Pel10]   R. Peled, On rough isometries of Poisson processes on the line. Ann. Appl. Probab. **20**(2), 462–494 (2010)

[Per99]   Y. Peres, Probability on trees: an introductory climb, in *Lectures on Probability Theory and Statistics*, Saint-Flour, 1997. Lecture Notes in Mathematics, vol. 1717 (Springer, Berlin, 1999), pp. 193–280

[Pet09]   G. Pete, *Probability and Geometry on Groups* (2009) (preprint)

[Pon86]   L.S. Pontryagin, *Selected Works*. Classics of Soviet Mathematics, vol. 2, 3rd edn. (Gordon & Breach Science, New York, 1986). Topological groups, Edited and with a preface by R.V. Gamkrelidze, Translated from the Russian and with a preface by Arlen Brown, With additional material translated by P.S.V. Naidu

[PSN00]   I. Pak, T. Smirnova-Nagnibeda, On non-uniqueness of percolation on nonamenable Cayley graphs. C. R. Acad. Sci. Paris Sér. I Math. **330**(6), 495–500 (2000)

[PZ32]   R. Paley, A. Zygmund, A note on analytic functions in the unit circle. Math. Proc. Cambridge Math. Soc. **28**(03), 266–272 (1932)

[Sch99]   R.H. Schonmann, Percolation in $\infty + 1$ dimensions at the uniqueness threshold, in *Perplexing Problems in Probability*. Progress in Probability, vol. 44 (Birkhäuser, Boston, 1999), pp. 53–67

[Sch01]   R.H. Schonmann, Multiplicity of phase transitions and mean-field criticality on highly non-amenable graphs. Commun. Math. Phys. **219**(2), 271–322 (2001)

[Shc13]   V. Shchur, A quantitative version of the Morse lemma and quasi-isometries fixing the ideal boundary. J. Funct. Anal. **264**(3), 815–836 (2013)

[Vir00]   B. Virág, Anchored expansion and random walk. Geom. Funct. Anal. **10**(6), 1588–1605 (2000)

[VSCC92] N. Th. Varopoulos, L. Saloff-Coste, T. Coulhon, *Analysis and Geometry on Groups*. Cambridge Tracts in Mathematics, vol. 100 (Cambridge University Press, Cambridge, 1992)

[Wat70]   M.E. Watkins, Connectivity of transitive graphs. J. Comb. Theory **8**, 23–29 (1970)

[Woe05]   W. Woess, Lamplighters, Diestel-Leader graphs, random walks, and harmonic functions. Comb. Probab. Comput. **14**(3), 415–433 (2005)

LECTURE NOTES IN MATHEMATICS  Springer

Edited by J.-M. Morel, B. Teissier; P.K. Maini

Editorial Policy (for the publication of monographs)

1. Lecture Notes aim to report new developments in all areas of mathematics and their applications - quickly, informally and at a high level. Mathematical texts analysing new developments in modelling and numerical simulation are welcome.

   Monograph manuscripts should be reasonably self-contained and rounded off. Thus they may, and often will, present not only results of the author but also related work by other people. They may be based on specialised lecture courses. Furthermore, the manuscripts should provide sufficient motivation, examples and applications. This clearly distinguishes Lecture Notes from journal articles or technical reports which normally are very concise. Articles intended for a journal but too long to be accepted by most journals, usually do not have this "lecture notes" character. For similar reasons it is unusual for doctoral theses to be accepted for the Lecture Notes series, though habilitation theses may be appropriate.

2. Manuscripts should be submitted either online at www.editorialmanager.com/lnm to Springer's mathematics editorial in Heidelberg, or to one of the series editors. In general, manuscripts will be sent out to 2 external referees for evaluation. If a decision cannot yet be reached on the basis of the first 2 reports, further referees may be contacted: The author will be informed of this. A final decision to publish can be made only on the basis of the complete manuscript, however a refereeing process leading to a preliminary decision can be based on a pre-final or incomplete manuscript. The strict minimum amount of material that will be considered should include a detailed outline describing the planned contents of each chapter, a bibliography and several sample chapters.

   Authors should be aware that incomplete or insufficiently close to final manuscripts almost always result in longer refereeing times and nevertheless unclear referees' recommendations, making further refereeing of a final draft necessary.

   Authors should also be aware that parallel submission of their manuscript to another publisher while under consideration for LNM will in general lead to immediate rejection.

3. Manuscripts should in general be submitted in English. Final manuscripts should contain at least 100 pages of mathematical text and should always include

   – a table of contents;
   – an informative introduction, with adequate motivation and perhaps some historical remarks: it should be accessible to a reader not intimately familiar with the topic treated;
   – a subject index: as a rule this is genuinely helpful for the reader.

   For evaluation purposes, manuscripts may be submitted in print or electronic form (print form is still preferred by most referees), in the latter case preferably as pdf- or zipped ps-files. Lecture Notes volumes are, as a rule, printed digitally from the authors' files. To ensure best results, authors are asked to use the LaTeX2e style files available from Springer's web-server at:

   ftp://ftp.springer.de/pub/tex/latex/svmonot1/ (for monographs) and
   ftp://ftp.springer.de/pub/tex/latex/svmultt1/ (for summer schools/tutorials).

Additional technical instructions, if necessary, are available on request from lnm@springer.com.

4. Careful preparation of the manuscripts will help keep production time short besides ensuring satisfactory appearance of the finished book in print and online. After acceptance of the manuscript authors will be asked to prepare the final LaTeX source files and also the corresponding dvi-, pdf- or zipped ps-file. The LaTeX source files are essential for producing the full-text online version of the book (see http://www.springerlink.com/openurl.asp?genre=journal&issn=0075-8434 for the existing online volumes of LNM). The actual production of a Lecture Notes volume takes approximately 12 weeks.

5. Authors receive a total of 50 free copies of their volume, but no royalties. They are entitled to a discount of 33.3 % on the price of Springer books purchased for their personal use, if ordering directly from Springer.

6. Commitment to publish is made by letter of intent rather than by signing a formal contract. Springer-Verlag secures the copyright for each volume. Authors are free to reuse material contained in their LNM volumes in later publications: a brief written (or e-mail) request for formal permission is sufficient.

Addresses:
Professor J.-M. Morel, CMLA,
École Normale Supérieure de Cachan,
61 Avenue du Président Wilson, 94235 Cachan Cedex, France
E-mail: morel@cmla.ens-cachan.fr

Professor B. Teissier, Institut Mathématique de Jussieu,
UMR 7586 du CNRS, Équipe "Géométrie et Dynamique",
175 rue du Chevaleret
75013 Paris, France
E-mail: teissier@math.jussieu.fr

For the "Mathematical Biosciences Subseries" of LNM:

Professor P. K. Maini, Center for Mathematical Biology,
Mathematical Institute, 24-29 St Giles,
Oxford OX1 3LP, UK
E-mail: maini@maths.ox.ac.uk

Springer, Mathematics Editorial, Tiergartenstr. 17,
69121 Heidelberg, Germany,
Tel.: +49 (6221) 4876-8259

Fax: +49 (6221) 4876-8259
E-mail: lnm@springer.com